b Institute

Interdiction Model for Exploring Halt Capabilities in a Large Scenario Space

Paul K. Davis, James H. Bigelow, Jimmie McEver

RAND

Prepared for the
Office of the Secretary of Defense

This report documents a relatively simple personal-computer model of interdiction in "the early halt problem"—i.e., the military problem of halting an invading army quickly. We anticipate that the report will be of interest to civilian and military analysts, to those supervising analysis, and to the modeling and simulation community because the report illustrates multiresolution design for exploratory analysis.

The model presented here (EXHALT—**ex**ploring the **halt** problem) stems from our conviction that many important defense-planning issues should be addressed—on a mission-by-mission or operation-by-operation basis—with relatively simple and focused models that permit broad-ranging exploratory analysis that confronts the uncertainty characterizing most of the key assumptions. Such analysis should not stand alone, but should instead be part of a family of analyses at different levels of detail and scope. Elsewhere, we describe related efforts with campaign-level models useful for integrative work and with high-resolution models that clarify underlying phenomena and the value of specific weapons or systems. Here, however, our emphasis is on a model for broad-ranging exploratory analysis at the aggregate level.

We have applied EXHALT to a Department of Defense (DoD) project on force transformation and joint experiments and in the Defense Science Board's 1998 summer study. Much of our emphasis involved understanding how command, control, reconnaissance, and surveillance capabilities affect the ability to achieve an early halt.

Our work was accomplished as part of a special cross-cutting project for the National Defense Research Institute's (NDRI's) Advisory Board (NAB). NDRI is one of RAND's federally funded research and development centers (FFRDCs). It is sponsored by the Office of the Secretary of Defense, Joint Staff, unified commands, and defense agencies. Comments are welcome and may be addressed to either McEver (Jimmie_McEver@rand.org) or Davis (Paul_Davis@rand.org) at RAND in Santa Monica, Calif.

CONTENTS

FIGURES

TABLES

EXHALT is a relatively simple, uncertainty-sensitive, and optionally stochastic treatment of the halt phase of an operation in which an attacking force (Red) is advancing on an objective while its armored vehicles are being interdicted by Blue, who uses joint long-range fires in the form of Air Force, Navy, and Army air power and of Navy and Army long-range missiles.[1] EXHALT is intended primarily for insight-oriented *exploratory* analysis. It is well suited to examining a broad scenario space (i.e., assumption space) because (1) it permits both parametric and probabilistic exploration (Davis and Hillestad, forthcoming); (2) it incorporates multiresolution modeling features, which allow users to work with more or less detail (Davis and Bigelow, 1998); and (3) it is interactive in somewhat the same way as a spreadsheet program. EXHALT is programmed in Analytica™, a visual-modeling system that runs on either a Windows™-based personal computer or a Macintosh computer. EXHALT is best used as one tool in a family of analyses that draws also upon, e.g., high-resolution simulations, campaign models, and empirical work, such as service and joint exercises (Davis, Bigelow, and McEver, 1999). Such a family is particularly important in understanding the synergism between ground-maneuver forces and long-range fires, interactions with allied forces, complexities of terrain and infrastructure, and weapon-level issues. That said, long-range fires

---

[1]Although EXHALT comes with a "default" data set of Blue Shooters and other scenario parameters of the type we have used in our analyses, the EXHALT structure allows the user to tailor Blue Shooter (and other) parameters to his or her analysis-specific needs.

are often a dominant factor, and EXHALT allows us to study many of their effects separately in a desktop model.

In addition to the usual inputs (e.g., sorties per day for aircraft), EXHALT includes aggregate-level situational factors that can affect outcomes. These relate to

- the *political-military scenario* (e.g., warning times)

- *strategies and tactics* (e.g., numbers of attack axes; vehicle spacing; and "asymmetric strategies" using, e.g., weapons of mass destruction or surprise)

- *force levels*

- *force effectiveness* (e.g., nominal kills per weapon and various measures of command and control, reconnaissance, and surveillance)

- qualitative *effectiveness factors* (e.g., the invader's "break point")

- certain aspects of the *environment,* such as terrain (e.g., size of open areas in which the invader can be targeted)

- *other analytical assumptions,* such as movement rate.

The conceptual model underlying EXHALT is straightforward. Red's advance begins on D-Day (treated as t = 0), perhaps with multiple axes and multiple columns per axis. Once the armored advance begins, Blue attacks Red's armored fighting vehicles (AFVs) (tanks, armored personnel carriers with weapons, etc.), but takes losses from Red's air defenses. To halt Red's movement, Blue must either destroy a specified fraction of Red's AFVs or employ a leading-edge attack that can in some cases stop Red's forward progress sooner. If the halt is not accomplished soon enough, Red reaches his objective. Alternatively, if Blue establishes a defense line, EXHALT can calculate whether Red reaches that line before being halted and, if so, what its residual force level is.

In attempting to halt Red, Blue begins with some forward-deployed forces. Depending on strategic warning, tactical warning, and access to regional bases, Blue may deploy additional forces and improve the readiness of both forces and command and control systems. Depending on his concerns about weapons of mass destruction—

particularly chemical and biological weapons—Blue may need to operate his forces from more distant locations and from less well-developed bases than he would prefer. This, in turn, reduces sortie rates for aircraft and the number of weapons per missile; it also reduces deployment rates and may affect the theater's effective capacity to accommodate and operate aircraft.

Blue's attacks on Red's AFVs are accomplished with only a fraction of Blue's shooters; the remainder are set aside for unsimulated attacks on, e.g., strategic targets and air defenses. The suppression of enemy air defenses is represented by an exponentially declining loss rate for Blue aircraft, which depends on the time to suppress air defenses. To account for the effectiveness of Red's air defenses during the early portion of the campaign, Blue flies only a fraction of his potential sorties for a "wait time" that is either specified as input or calculated by a commander model.

At the beginning of each day, an optional submodel representing Blue's commander (i.e., a "Blue agent") decides what fraction of manned aircraft to use and whether to use certain high-value, long-range missiles that depend on command and control assets that may not yet be at full effectiveness. Use of manned aircraft is unrestrained after the wait time is complete and after reconnaissance, surveillance, tracking and acquisition assets are able to provide high-quality information. Before then, the Blue commander makes trade-offs between minimizing the enemy's penetration into friendly territory and minimizing his own casualties due to Red's air defenses. The Blue commander sets the wait time accordingly. A variant of the commander model could trade-off Red force levels expected to reach a Blue defense line against Blue's casualties to air defenses. The trade-off relationships are, of course, inputs, since they reflect alternative strategies and values.

During each time step, Blue gains newly deployed shooters, decides how many shooters to employ against Red's advancing armor columns, attacks the columns, and takes losses. Red's losses are calculated, and his new position is determined. If Red reaches his objective or the postulated defense line, Red AFVs accumulate there (Red does not continue to advance beyond the objective, and close combat is not simulated, although we have sometimes added a crude representation). If the Red force is halted, his air defenses are assumed to be defeated as well, and Blue continues his attacks (taking no further losses) until the Red force is annihilated.

The effectiveness of Blue attacks depends on many factors, including command, control, communications, computers, intelligence, surveillance, and reconnaissance (C^4ISR) factors and Blue's force-employment strategy. Indeed, a major virtue of EXHALT is in highlighting Blue's *system problem*: His effectiveness in bringing about an early halt can be greatly reduced by any of many shortcomings (Davis, Bigelow, and McEver, 1999). In particular, an early halt will often require having a large number of shooters in place by D-Day, having highly capable long-range fires, having survivable C^4ISR systems operating from the outset of war, being able to operate shooters before air defenses are fully suppressed, and slowing Red's advance rate with fires focused on Red's leading-edge units or in ways EXHALT itself does not model (e.g., with bombing and special operations that create roadblocks and logjams, or with allied ground-force units that can block, delay, or at least harass advancing units).

EXHALT can be extended or simplified readily because of its modular design and the character of the Analytica environment. Although we have provided considerable explicit flexibility—including switches to turn model features on and off and two user modes at two different levels of detail—users who are able to program may well want to change the model itself, not just the data. EXHALT can also be a module in a larger system model. Alternatively, detail can be added to EXHALT by replacing a given input variable with a function generating that input.

Finally, we emphasize that EXHALT can be used for either parametric or probabilistic exploratory analysis to assess the effects of uncertainty, including stochastic processes (Davis and Hillestad, forthcoming). Indeed, this is one of EXHALT's principal features.

ACKNOWLEDGMENTS

We acknowledge the assistance of colleague Manuel Carrillo, who developed several software tools that facilitated our development and use of EXHALT. We also benefited from discussions of the halt problem with Glenn Kent, Ted Harshberger, John Gordon, and Richard Hillestad.

ABBREVIATIONS

AFV	Armored fighting vehicle
APC/T	Armored personnel carrier/transport
ATACMS	Army Tactical Missile System
BAT	Brilliant anti-armor submunition
C^2	Command and control
C^4ISR	Command, control, communications, computers, intelligence, surveillance, and reconnaissance
EXHALT	EXploratory analysis of the HALT problem
JICM	Joint Integrated Contingency Model
MRM	Multiresolution modeling
N/TACMS	Naval Tactical Missile System
RSTA	Reconnaissance, surveillance, targeting and acquisition
SEAD	Suppression of enemy air defenses
SFW	Sensor-fused weapon
START	Simplified Tool for Analysis of Regional Threats
WMD	Weapons of mass destruction

Chapter One

INTRODUCTION

PURPOSE

This report documents a model called EXHALT (an acronym for "exploring the **halt** problem"). The model is programmed in a visual-programming system called Analytica™.[1] Analytica permits user-friendly modeling that supports both parametric and probabilistic exploration of uncertainty, including uncertainty related to random (stochastic) processes in a given simulation. Because Analytica models can be relatively comprehensible and self-documenting, this report is less detailed than traditional documentation. Our emphasis is on the conceptual model, its structure, its inputs and outputs, and some illustrative displays. Readers needing more-precise information should examine EXHALT itself. Readers who may be unfamiliar with the Analytica modeling environment are strongly encouraged to work through the Analytica Tutorial before tackling EXHALT.

MODEL OBJECTIVES

While explicitly dealing with large numbers of uncertainties, the objectives of this model and related analysis are

---

[1]Analytica was originally a product of Carnegie-Mellon University and is now being distributed, maintained, and extended by Lumina Decision Systems, Inc. Its web site (http://www.lumina.com) explains options for free testing of the software and existing Analytica programs. We used Release 1.0.1 for the Macintosh, which is well documented in Lumina (1996). EXHALT can be used interchangeably on Macintosh or Windows-based versions of Analytica.

1. to observe the effects of various factors (e.g., platforms and weapons; suppression of enemy air defenses [SEAD] effectiveness; invader and interdictor tactics; and command, control, communications, computers, intelligence, surveillance, and reconnaissance [C⁴ISR]) on the success of air and missile attacks on an advancing armored column

2. to characterize regions of the scenario space (the space of all assumptions) by whether Blue's "halt capability" is good, marginal, or poor

3. to identify potential improvement measures involving new systems, doctrine, or actions during warning

4. to demonstrate hierarchical and multiresolution modeling principles that facilitate exploratory analysis.[2]

EXHALT's objectives are narrow. For example, it does not include close combat, although adding a representation of that would be quite easy. The reason for not doing so is that this model is intended to address only one "module" of the larger halt problem (which is, in turn, one module of an overarching campaign) in a relatively simple way. Rather than include a simplistic representation of close combat (e.g., something amounting to a Lanchester-equation depiction), we prefer to examine such combined-arms issues in a campaign-level model, such as the Joint Integrated Contingency Model (JICM),[3] which can represent road networks; terrain; nonlinear combat, such as turning movements and rear-area operations; allied forces; and details of strategic and tactical mobility. There one can readily "see" how long-range fires, maneuver forces, mobility, and other factors

---

[2]This builds on ideas laid out in Davis and Bigelow (1998) and Davis and Hillestad (forthcoming).

[3]JICM is a global warfare model that includes strategic and tactical mobility and description of regional conflicts at the operational level of detail. It has been used extensively in studies for the Office of the Secretary of Defense (OSD), the Joint Staff, Air Force, and Army. War colleges have been using JICM for some years as, more recently, have the DoD's Office of Program Analysis and Evaluation (PA&E) and the Republic of Korea. A related spreadsheet model, the Simplified Tool for Analysis of Regional Threats (START) uses a simplified depiction of theater-level maneuver, but also incorporates a joint perspective. Developed by Bruce Bennett and Barry Wilson, it has been used primarily for the Office of Net Assessment and the Air Force.

interact, but at the expense of greater complexity.[4] Again, the purpose of EXHALT is to focus on just the issue of fires.

More generally, EXHALT was developed as part of a family-of-analysis, family-of-models approach that draws on a diversity of model-based and experimental work and that recommends priorities for service and joint experiments (Davis, Bigelow, and McEver, 1999). It was also used in the 1998 Defense Science Board summer study to show the potential impact of plausible advanced systems and in a recent study of future ground forces for rapid intervention (Gritton et al., 2000).[5]

STRUCTURE OF THE REPORT

With this background, we proceed in Chapter Two to give a relatively more detailed description of EXHALT. Chapter Three then discusses inputs and outputs for two modes of operation differing sharply in level of resolution. Chapter Three also shows illustrative displays of exploratory analysis. Appendix A provides more detail on the "commander model," and Appendix B provides a data dictionary. As noted earlier, we expect that users will rely primarily on EXHALT's self-documenting features, except for a single reading of this document. Appendix C addresses verification and validation issues regarding EXHALT. Appendix D provides a brief description of a "batch-run" tool that can assist users in working around computer memory problems associated with spanning a scenario space with many degrees of freedom. Appendix E illustrates use of Analytica's visual-modeling environment; it shows some lowest-level computer code and how to change it.

---

[4]For a 1996 study of the halt problem at the joint and combined campaign level, see Davis, Schwabe, Nardulli, and Nordin (forthcoming).

[5]Others have also begun to emphasize the value of personal-computer models and families of the sort we have been recommending. For an account of some related models developed at the MITRE Corporation, see Belldina et al. (1997). See also Ochmanek et al. (unpublished manuscript).

Chapter Two

THE CONCEPTUAL MODEL UNDERLYING EXHALT

TOP-LEVEL VIEW

Figure 2.1 describes the top-level flow of EXHALT. Figure 2.1 should be seen as describing data flow at a particular time step—i.e., it applies at an arbitrary snapshot in time. Arrows show the direction of data flow; dashed arrows indicate where data from one node at the previous time step is used as input to another node in the present time step (i.e., feedback). It is implicit that each variable's update during the time step may depend on its own values at previous times. Subsequent figures (which are copied directly from the computer screen) include what appear to be double-headed arrows. These are actually a combination of a solid forward arrow and a mostly obscured dashed feedback arrow. We shall describe Figure 2.1 briefly and then walk through each of the modules shown in somewhat more detail.

The **Update Blue Shooters** module takes information regarding Blue's forward-deployed aircraft and missiles, his warning time (tactical and strategic), deployment rates (for various types of shooters), losses taken in the previous period, effectiveness status as of the previous period, and deployment or theater capacity limits to calculate the number of "shooters" of each type Blue has available in the theater at a given time. How Blue uses these shooters is determined in another module.[1]

---

[1]The variable Shooters, like most other variables in EXHALT, is described mathematically by an array—in this case, a simple vector with a component for each type of shooter (e.g., F-15E/F, the Army Tactical Missile System [ATACMS], AH-64). Another

RAND *MR1137-2.1*

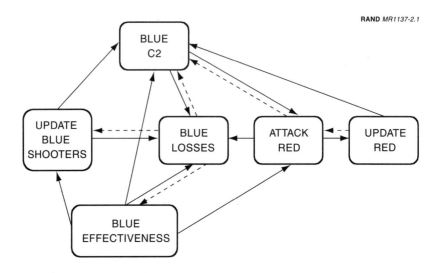

Figure 2.1—Top Level of EXHALT

The **Blue Effectiveness** module uses inputs for Blue's C⁴ISR capabilities, kill and sortie rates for Blue's various shooter types, and some possible strategies Red and Blue might use (e.g., Red may use dash-and-hide tactics, and Blue may concentrate attacks on Red's leading-edge armored fighting vehicles [AFVs]) to determine the effectiveness of Blue's sorties against Red's armor. The tactics and effects are discussed in more detail later.

The **Blue Losses** module calculates the shooter losses that Blue takes during the halt campaign, using Blue shooter and sortie tallies, and a loss-per-sortie or -shot function that declines over time as Red's air defenses are degraded.

The **Blue Command and Control (C2)** module then allocates some fraction of each type of Blue's available shooters to anti-armor missions, withholding some for other missions, such as strategic attacks; it may also withhold some of Blue's shooters if Blue is still concerned

---

array, Blue Shooter Inputs, is indexed by Shooters and by Shooter Characteristics to create a table (a two-dimensional array) containing all initial Blue shooter data. Subsequent nodes within EXHALT "reach back" to this input matrix to derive such vectors as Arrival Rate and Initial Shooters in Theater (both indexed by Shooters).

about strong Red air defenses, or if C[4]ISR assets are not yet adequately effective. To accomplish this, the Blue C2 module can employ two optional "decision agents" (so called because they represent, i.e., act as agents for, human decisionmakers), which are described in more detail in subsequent sections.

In the **Attack Red** module, Blue's shooters attack Red's armor. The Attack Red module calculates Red's losses and the number of AFVs remaining in Red's advance. Finally, the **Update Red** module updates Red's position and numbers, making use of Red's speed of advance. The next wave of Blue attacks then begins, and the model updates Blue's numbers for losses taken in the last wave and new deployments that may have arrived.

Let us now describe each of the above modules in somewhat more detail. Even more detail is provided within the model itself. Indeed, we anticipate that—except for one reading of this overview report— users of EXHALT will rely almost entirely upon the model's self-documenting features.

UPDATE BLUE SHOOTERS

This module, the top-level diagram for which is shown in Figure 2.2, calculates the number of shooters of each type present in the theater through the halt campaign.[2] Initial shooters, arrival rates, deployment caps, and theater capacities are read from the Blue Shooter Inputs matrix (described in Appendix A). Arrival rates and theater and deployment shooter limits can further be affected by a possible Red threat of the use of weapons of mass destruction (WMD), Red mining of waterways, or the ability of Red (and his allies) to constrain Blue's ability to use bases convenient to the theater. These factors, along with the tactical warning time, are used to calculate the

---

[2]The conventions for this and all similar drawings are as follows: Deterministic inputs are indicated by rectangular or trapezoidal nodes (shown in later diagrams). Inputs determined by random draw are indicated by oval nodes. Outputs are indicated by hexagonal nodes (these will appear in subsequent figures). Single-variable nodes are indicated by oblongs, while modules containing two or more nodes are indicated by oblongs with darker edges and bold lettering. Italic lettering indicates that the node shown is being copied here for convenience (as an "alias"), but is actually located in another module. Tiny arrows to the immediate left or right of a node indicate influences from or on nodes outside the current diagram.

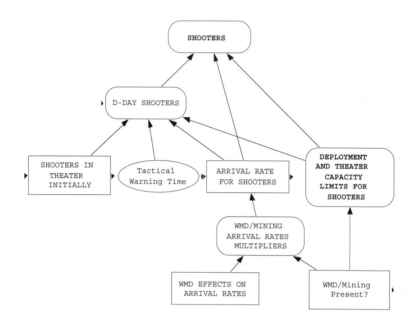

Figure 2.2—Top Level of the "Update Blue Shooters" Module

number of shooters of each type present in the theater when the Red advance begins (D-Day). During each step, new shooters arriving in the theater are added (if theater air base capacity has not been reached), and the losses taken from the last step by Blue from Red's air defenses are subtracted. The resulting vector of available Blue shooters, indexed by shooter type, is then passed on to the C2 module, in which some portion of the shooters will be allocated to anti-armor missions.[3]

Figure 2.2 identifies two input nodes, those for tactical warning time and for the presence of WMD or mining, as probabilistic input variables.[4] These nodes take advantage of Analytica's ability to assign

---

[3]In fact, this information is also passed to the Blue Losses module, as indicated by Figure 2.1.

[4]EXHALT is a stochastic model, but we reserve the term *stochastic* for instances in which we are referring to processes that are random within a given real or simulated war, as distinct from parameters that may have well-defined values for a given war but

input values drawn from probability tables or distributions specified by the user. Importantly, this feature allows us to use EXHALT to examine some of the real effects of uncertainty in the scenario space of the early halt problem.

BLUE EFFECTIVENESS

This module calculates the effectiveness of Blue's attacks on Red's AFVs. The top-level Analytica diagram for this module is shown as Figure 2.3. It includes several time-dependent multipliers to Blue's nominal effectiveness. An "Effectiveness Multiplier Source?" switch allows users either to input the aggregate multiplier to Blue's effectiveness (indexed by shooter type) or to have it calculated from higher-resolution factors. Each of these factors is merely driven by parameters in EXHALT, but the parameter ranges and uncertainty distributions used should be based on higher-resolution analysis and empirical work (Davis, Bigelow, and McEver, 1999). In any case, the factors treated here are as follows:

- *Attack-mode effects due to Blue's attack strategy.* Blue can attack with either a leading-edge or in-depth strategy. The leading-edge strategy (Ochmanek et al., 1998) allows "rollbacks" in that, if Blue is able to completely shatter the forward x km of a column on a given day, the column's effective advance will be reduced by x km/day. If Blue has enough interdiction capability, x may be greater than the nominal column speed, V, in which case the advance will be rolled back by (x–V) km/day.[5] Further, since leading-edge attacks are concentrated on forward units, Blue can stop an entire unit (i.e., remove that unit from the advance) by destroying some fraction of that unit, specified as the "Unit

---

that we must treat as uncertain in advance of the war itself (Davis and Hillestad, forthcoming). To illustrate, the sortie rates generated on a given day during an actual war might be a stochastic variable from day to day, while the mean sortie rate we use in analyses of future wars is better seen as an uncertain parameter that would vary from war to war.

[5]The term *rollback* is not rigorous, since, in principle, a portion of the force in the "rollback area" could still take cover ("go to ground") and mount opposition, rather than flee.

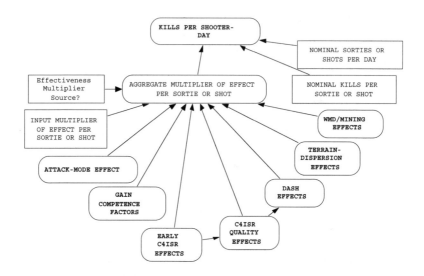

Figure 2.3—Top Level of the "Blue Effectiveness" Module

Break Point."[6]   We assume that Blue suffers an effectiveness degrade in these leading-edge attacks, since concentrating attacks at Red's front is more difficult than allowing attacks throughout Red's advance. Blue's in-depth strategy suffers no such degrade but likewise has none of the other advantages of the leading-edge strategy; in particular, the advance continues to move at a constant rate until the "halt fraction" of his forces has been killed, and the advancing units suffer no "unit dissolution" effects, since Blue's attacks are distributed throughout the advance. The leading-edge strategy's advantages, which can be dramatic, diminish quickly as the number of axes of advance—or the number of columns per axis—increases (Davis and Carrillo,

---

[6]This effect is like the larger "Halt fraction" effect, by which the entire advance falls apart after some fraction of the advance is destroyed. Use of the Unit Break Point assumes that a unit will fall apart after some fraction of its component armor is destroyed, removing all of that unit's AFVs from the advance. (The Unit Break Point can be set to 1 if users wish to employ the leading-edge rollback effect, but not this "unit dissolution" effect.) The effect of the Unit Break Point appears in the Attack Red module.

1997), although such a multiaxis advance might also be slower if the extra roads were poorer than the main road (Ochmanek et al., 1998).

- *Gain competence factors.* Unless given adequate tactical warning time or exceptional training to allow Blue to exercise his C⁴ISR personnel and assets, Blue will not begin the campaign with his C⁴ISR assets fully effective. This module calculates a time-dependent multiplier to Blue effectiveness due to this factor. With time, Blue's competence in command and control matters approaches its nominal level.

- *Early C⁴ISR effects.* Blue experiences time-dependent multipliers to his effectiveness that are related to the early deployability and survivability of his C⁴ISR system. Blue's effectiveness is degraded early, depending on how effective Blue's C⁴ISR assets can be in the face of Red's air defenses.

- *C⁴ISR quality effects.* These effects are represented parametrically in the C⁴ISR Engagement Factor, which accounts for such effects as the ability of Blue to provide targets to his shooters quickly; his ability to see through foliage to acquire and hit targets; and the saturation point of his reconnaissance, surveillance, targeting, and acquisition (RSTA) system. This factor has qualitative values like "Base" or "Enhanced" that represent different levels of Blue C⁴ISR system capability. Characteristics of the several levels are based loosely on results from high-resolution studies of potential capabilities.

- *Dash effects.* As a tactic to counter Blue's attacks, Red can—in theaters that have appropriate hiding places—adopt "dash" tactics, in which he concentrates his movement during relatively small parts of the day, then goes into hiding the rest of the day (Davis and Carrillo, 1997). Red is constrained in doing so by his ability to coordinate such movement, the terrain, and other factors. Blue can react to this strategy by concentrating his attacks during Red's movement periods or by having systems capable of finding and attacking hiding Red AFVs. Although this is currently a parametric calculation to make a point, the module could be replaced in the future by a more phenomenological representation based on higher-resolution analysis. For example, one response to this Red countermeasure would be Blue combat air patrol stations.[7]

- *WMD and mining effects.* If Red can threaten to use chemical, biological, or even nuclear weapons or to mine critical waterways, Blue's ability to deploy to the theater and attack can be slowed. This module highlights this factor (although the effects of WMD are actually distributed across different parts of the model).[8]

- *Terrain and dispersion effects.* The presence of mixed terrain, in which Red is only vulnerable when in open areas of the terrain, reduces the effectiveness of Blue's attacks. This effect is complicated by the fact that Red AFVs may travel in platoon and company packets, meaning that Red's presence in open areas is not uniform. This module allows users to input a shooter-dependent multiplier to represent this effect. A more detailed model, Reproduction of the Precision Engagement Model (RPEM), is provided "on-line" in the Terrain/Dispersion Effects module to assist users in arriving at reasonable multipliers. RPEM is derived from a higher-resolution model developed from entity-level simulation data.[9]

These multipliers are combined with the nominal values of kills per sortie or shot and sorties per day for each shooter type to determine the effectiveness of each of Blue's shooter types at any point in the campaign.[10] These effectivenesses are then reexpressed as multiples

---

[7]Colleague Glenn A. Kent has studied some possibilities in this regard.

[8]The threat of Red's use of WMD and/or mining affects (in EXHALT) Blue's sortie rates, arrival rates, and theater shooter capacity.

[9]This factor is calibrated to a more detailed stochastic model (Precision Engagement Model [PEM]), which in turn was motivated by and calibrated to results of high-resolution simulation (Davis, Bigelow, and McEver, forthcoming). RPEM is not connected to EXHALT but can be used to generate estimates of terrain-dispersion effects for various weapon systems in a given scenario, so that users can insert reasonable multipliers into the terrain-dispersion effects input vector.

[10]In the current version of EXHALT, the nominal value of kills per sortie or shot of a given platform does not depend explicitly on weapon loading or the spacing of vehicles. To use EXHALT in a weapon-mix study, however, one would need to include these dependences explicitly. See Ochmanek et al. (unpublished manuscript) for a game-theoretic discussion of why Blue having a mix of weapons could limit the value of Red's obvious tactic of dispersing vehicles to reduce the effectiveness of Blue's area

of a "standard shooter's" effectiveness.[11] This information is passed to the Blue C2 module, in which the shooters-available vector is converted to a simple number of equivalent shooters (i.e., a scalar rather than a vector) and assigned to anti-armor attacks. This is convenient, for reasons discussed later, but does *not* reduce generality.

BLUE C2

EXHALT contains a "commander model" (Figure 2.4), which takes the shooter vector produced in the "Update Blue Shooters" module and—accounting for withheld sorties due to the wait time, attacks on strategic and other targets, and the decision to withhold missiles until early effectiveness degrades have been mitigated—generates the number of "equivalent shooters" that will attack Red's armored targets.

The C2 module also includes decision models dealing with strategy and tactics. These are simple "agent" models that represent human commanders faced with imperfect information and difficult trade-offs. The missile-launch decision is a rather trivial example of such a decision agent, one allowing Blue to withhold the launching of his missiles until C[4]ISR and other factors are sufficient to ensure a threshold level of effectiveness, rather than simply scripting the missile launch for some predetermined time that would be inappropriate if other parameters, such as Competence Time from Warning or C[4]ISR System, were changed.

A more intricate example of a decision agent is the optional wait-time decision module. In this, a Blue commander estimates various campaign outcomes as a function of his choices and, based on those estimates, chooses a "wait time" (how long Blue will fly at a reduced rate because of relatively robust Red air defenses at the beginning of the campaign), that best matches his criteria for success. This involves trading off likely Red penetration distance against likely losses of Blue's air forces.[12] Once the wait-time decision agent selects

---

weapons. See Davis, Bigelow, and McEver (forthcoming) for discussion of how dispersal, weapon characteristics, terrain, and other factors interact.

[11]The default, which is easy to change, is that the F-15E is a standard shooter.

[12]Since this is a unique and fairly complex feature of the EXHALT model, it is described in some detail in Appendix B.

the best wait time,[13] any wait-time sortie effects and set-asides for attacks on strategic targets (i.e., targets other than Red AFVs) are accounted for.

Each shooter type in the resulting vector is then converted into equivalent shooters by multiplying it by its relative effectiveness (in terms of kills per shooter-day) compared to the effectiveness of the "standard" shooter. The components of the vector are then summed to produce a scalar value of equivalent shooters that will attack Red AFVs during the current step. This step is mathematically unnecessary and introduces no further approximations, but, in keeping with the multiresolution design of EXHALT, it permits (without requiring) users to simplify the model by scripting the attacks of equivalent shooters, as is sometimes convenient.

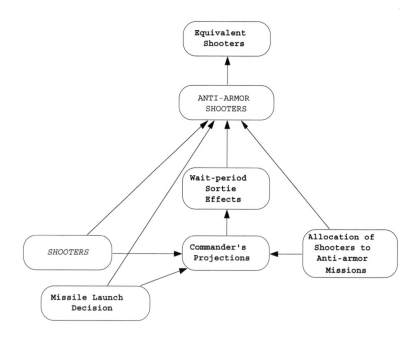

Figure 2.4—Top Level of the "Blue C2" (or "Blue Commander") Module

---

[13]As described in Chapter Three, the wait-time decision agent can be switched off, in which case a user-specified input value for the wait time will be used.

BLUE LOSSES

This module (Figure 2.5) calculates the losses that Blue takes during the Halt campaign. Each shooter type has, as an input, an initial loss rate. This rate is defined as the probability that a shooter of that type will be lost to Red air defenses or other factors during a given sortie.[14] Each shooter's loss rate declines exponentially at a rate based on the SEAD time. Since the Halt point is considered to be the point at which Red's advance collapses, Red's air defenses are consid-

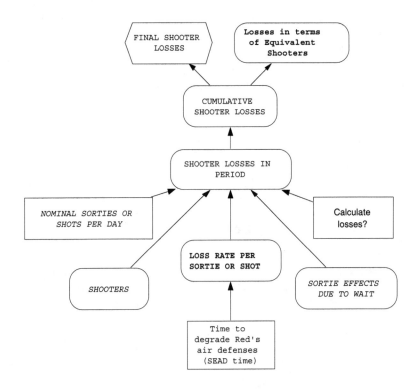

Figure 2.5—Top Level of the "Blue Losses" Module

---

[14]In the baseline version of the model, only aircraft losses are accounted for, but one can enter loss rates for missile batteries as well; those would be loss rates per shot rather than per sortie. They might be the result of counterfire or special operations forces.

ered to collapse there as well and become inactive once halt is achieved.

The module calculates the number of sorties flown in a given time step by each shooter type and then calculates the losses to each shooter type based on that type's loss rate at that time. Shooters are assumed to have carried out their attacks before they are lost, and the losses are subtracted from the shooters available for the next step's attacks.

To provide a single number as a measure of overall losses, part of the module converts the vector of lost shooters to a single number of equivalent shooters. However, the full vector of losses by shooter type is used in determining the number of shooters available. Should such detail not be needed, users may switch off the Blue losses calculation (which also turns off the wait-time decision agent, since one of the agent's decision criteria is Blue losses).

ATTACK RED

This module (Figure 2.6) calculates the number of Red armored vehicles that Blue kills in a given time step by multiplying the number of equivalent Blue shooters attacking Red armored targets by the kills per standard shooter-day of the equivalent Blue shooter, and then adjusting for the length of the time stamp.

If Blue uses the "leading-edge" attack strategy, additional Red armored vehicles are considered to be disabled by Blue's attacks. As mentioned in the explanation of the leading-edge attack strategy in the "Blue Effectiveness" section, we reason that Blue's leading-edge attacks are concentrated on a particular segment of the Red advance and that each unit within that segment has a "break point" beyond which the remainder of the attacked units cannot continue to function. The "Unit Break Point" is specified as an input, and once that fraction is destroyed, the remaining AFVs in that unit are disabled and removed from the advance.[15] In EXHALT, the sum of AFVs disabled in this way and those directly killed are referred to as "AFVs Stopped."

---

[15]We assume that Blue is able to allocate and target his sorties so that he destroys each Red unit with the minimum force possible.

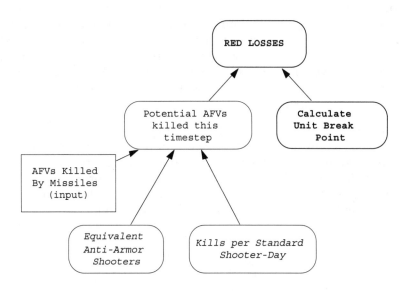

Figure 2.6—Top Level of the "Attack Red" Module

Once Blue's potential to destroy or disable Red AFVs in the current time step is determined, this potential is compared to the number of Red AFVs remaining, to ensure the Blue cannot stop more AFVs than remain.

UPDATE RED

This module (Figure 2.7) updates the position of Red's advance as a result of this period's movement (if any) and attacks. If the Overall Halt Fraction of Red's initial armored force has not been destroyed or disabled, Red will advance at his base movement rate (adjusted for the length of the time step), minus any "rollback" that may have occurred as a result of Blue's attacks. Rollback calculations are detailed in the "Calculate Length of Column Stopped by Blue Attacks" module within the "Calculate Red Movement and Position" module.

If Red's position places him at the objective, his penetration will stop there, but Red armored vehicles will begin to accumulate at the objective at a rate determined by the base movement rate, the roll-back, the spacing, and the number of axes and columns along which Red advances. Red forces will accumulate at the objective until all remaining Red forces arrive there; all remaining Red forces have been destroyed or disabled; or Blue's rollback of Red is sufficient to push him back from the objective (in which case, the Red armor at the objective remains there).

If more than the Overall Halt Fraction of Red's initial armored force has been stopped, Red's base movement rate drops to zero, and the only movement that takes place is due to Blue's rollback of Red forces (although, in general, the distance measurement most of interest is the distance of maximum Red penetration, which will _occur while Red has a positive base movement rate).

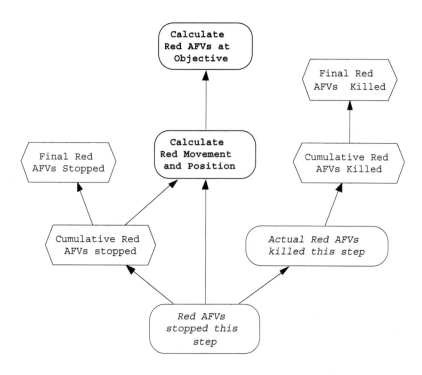

Figure 2.7—Top Level of the "Update Red" Module

The module also keeps track of the cumulative number of Red armored vehicles stopped and destroyed, as well as marking how many total Red armored vehicles were stopped and destroyed at the end of the simulation. The user can also specify an "Interim Point," some distance along Red's advance, and observe Red's strength, in terms of number and fraction of AFVs remaining. This is valuable for the purpose of considering possible defense lines by small Blue ground forces (including allies). Rather than actually modeling close combat, one can simply observe the strength that would be needed at the defense line, when Red arrives there, to achieve the desired force ratio.

Chapter Three

INPUTS AND OUTPUTS OF EXHALT

An advantage of running a model, such as EXHALT, in the Analytica environment is the ease with which the inputs and outputs of the model are customized to suit the user's purposes. Since EXHALT was constructed with a multiresolution design, users can move up and down in levels of resolution when using EXHALT to study the halt problem. In some cases, the options are built in; in others, the user would need to modify the model itself a bit, but that is usually easy and does not require the kind of programming expertise needed for changes in C programs, for example.[1]

MULTIRESOLUTION MODES OF EXHALT

EXHALT is intended to be a dynamic model; that is, once users understand the logic and structure of the model, they should be able to manipulate nodes and modules, taking advantage of Analytica's ease of use, to have EXHALT represent the halt problem at the proper level of resolution and with the proper variables for a particular set of needs. For illustrative purposes, we have constructed a "truncated" resolution mode so that users can realize the flexibility and utility of the multiresolution modeling (MRM) design of EXHALT. What follows is a discussion of each of the resolution modes and the inputs each requires.

---

[1]Some users may be used to thinking of "good" models as models with nothing "hard wired" and everything of interest changeable in "data." That image is no longer as appropriate when dealing with high-level interactive model environments, such as those of Analytica or Excel. Changing "the model" is usually quite simple.

Resolution Mode 1: The "Full" EXHALT Model

Resolution Mode 1 is the most detailed level of resolution developed in this project. Mode 1 is characterized as follows:

- Blue's shooters are represented by a vector indexed by shooter type, each of which can (among other things) deploy at different rates, attack with different sortie and kill rates, possess type-specific loss-rates, and be allocated to anti-armor missions differently.

- Blue's shooters may include, in addition to aircraft, missile batteries, which fire against Red AFVs only after a Blue decision agent gives the go-ahead for launch.

- Blue's wait time is also decided by a decision agent, which optimizes the wait-time decision based on Blue's projections for Red penetration distance and Blue casualties and his utility trade-off between the two. As mentioned earlier, this agent can be switched off to allow the wait time to be set as an input.

- Blue must choose an attack mode that, depending on the choice, can slow or even reverse Red's advance by concentrating shooter and missile attacks on the front of Red's column.

- Shooter effectiveness is predicted by a host of factors, including C^4ISR, terrain, WMD, Red march strategy, and Blue attack mode effects (some of which may affect each shooter type differently). A switch turns off this calculation to allow users to input the aggregate multiplier to effectiveness directly.[2]

Full information concerning the input domain of Resolution Mode 1 is provided in Table 3.1. A complete input data dictionary describing all input variables to the full EXHALT model is attached as Appendix B.[3]

---

[2]Note that, even when the user switches off the more-detailed calculation, the user must still specify which attack mode is used and whether Blue is subject to a WMD or mining threat. These factors affect the rollback of Red's advance and deployment rates and caps (respectively) in addition to effectiveness.

[3]Users may wish to separate the input-data file from the model so as to maintain a variety of base cases or to maintain personal databases. Defining the interface module as, essentially, a library module makes this straightforward.

Table 3.1

Input Parameters for EXHALT, Resolution Mode 1

Blue Inputs	
Blue Shooter Input Matrix	
Shooter Characteristics	
Shooters in theater initially	Nominal kills per shot or sortie
Arrival rate	Fraction flying during wait period
Maximum deployment	Nominal anti-armor fraction
Shooter capacity of theater	Initial shooter loss rate
Nominal shots or sorties per day	Missile or aircraft?
Shooter Types	
F-22 equivalents	Arsenal aircraft
F/A-18 E/F equivalents	Naval missiles
B-1 equivalents	ATACMS (BATs)
F-15 E equivalents	NTACMS (Tubes)
Other parameters for Blue	
Mode of Blue attack	C^4ISR system
SEAD time	Terrain-dispersion multiplier
Flexibility of fires	Missile effectiveness threshold
CVBG arrival time	Wait-time decision criteria
Competence time from warning	

Red Inputs	
Red divisions	Mean spacing between Red AFVs
Red AFVs per division	WMD/mining flag
Axes of Red advance	Red time concentration factor
Columns within each axis	Distance to objective
Base Red column speed	

Model Assumptions	
Unit break point	System Inputs
Overall halt fraction	Max time
Tactical warning time	Time step
Strategic/tactical Δt	Interim point

In a way, labeling Resolution Mode 1 as the "full" model is a misnomer. Resolution Mode 1, the highest-resolution mode we developed, can easily be expanded to model shooter weapon types (e.g., area weapons, one-on-one missiles, brilliant anti-armor submunitions [BATs,]) and payload mixes. Such embellishment is left to users requiring this level of resolution. Examples of such detail can be found in Davis, Bigelow, and McEver (2000) or Ochmanek et al. (unpublished manuscript).

Several "switches"—nodes that allow users to turn off Blue loss calculations, the wait-time decision agent, and the calculation of Blue's effectiveness multiplier from higher-resolution factors—add to EXHALT's flexibility. Note that, since the wait-time decision agent requires input from the loss calculation, when losses are turned off, the wait-time decision agent must also be turned off. These switches were created to facilitate cases in which, for example, a user might want to explore the effect of the wait-time parameter and would therefore provide it exogenously rather than having it determined by a decision agent.[4] When a switch is used to disconnect the wait-time decision agent or the higher-resolution calculation of Blue effectiveness multipliers, the model will draw on user-specified input values to replace these calculations. If Blue losses are turned off, any casualties Blue may suffer are ignored.[5]

Resolution Mode 2: The Truncated Model

The need for a truncated version of EXHALT arose during an exploratory study of the halt problem (Davis, Bigelow, and McEver, 1999). The full EXHALT model is simple compared to high-resolution simulation but still has far too many degrees of freedom for exploring broad regions of the scenario space. Thus, we can take advantage of EXHALT's MRM design to arrive at a much simpler model (similar to one used in Davis and Bigelow, 1998). The degree of "pruning" required for any particular exploratory study will vary depending on the focus of the study. Resolution Mode 2 is intended to be an example of how EXHALT can be radically truncated for use at a much lower level of resolution.

In Mode 2, all information about Blue shooter inventories, including deployment, losses, theater capacity, diversity of shooter types, anti-

---

[4]The switches can also be used for cases in which computer RAM is constrained. For example, implementation of the wait-time decision agent adds a rather complex calculation to the model. If decision-agent factors are not key in a particular study, the appropriate switch can be used to remove the agent from the calculation.

[5]Currently, EXHALT is implemented such that the wait-time decision agent is automatically disabled when the Blue loss calculation is disabled. Users should note, though, that EXHALT will use the input value for wait time when the agent is off. Thus, users should ensure that the input wait time is reasonable for their analysis when switching off the agent either directly or indirectly through turning off the loss calculation.

armor fraction, etc., is condensed into the scalar Blue Shooters inputs. This is a time-indexed vector, into which the user enters a "script" of the number of equivalent shooters attacking Red AFVs each day. No generality is lost, but the script entered implicitly includes the effects of Blue losses, mission assignments, deployment rates and limits, WMD effects, etc. So, if losses are not incorporated into the script, the truncated model becomes a "no-loss" model. In the simplest case, the scripts simply use constant deployment rates.

Resolution Mode 2 has other differences from Resolution Mode 1, as well. The effectiveness of Blue's equivalent shooter type is an input parameter, rather than a calculation from higher-resolution variables. Blue's missiles can either be turned into equivalent shooters or treated separately, with a single one-time subtraction from Red's forces at the beginning of the campaign (e.g., the result of the use of missiles early in the conflict). The following is a full listing of Mode 2's input parameters:

- Anti-armor equivalent shooters (time scripted)
- Effectiveness of Blue equivalent shooter
- AFVs killed by Blue's missiles
- Initial Red AFVs
- Mean velocity of Red advance
- Distance to objective
- Overall halt fraction
- Unit break point
- Maximum time
- Time step
- Interim point.

Figure 3.1 illustrates the data flow of the truncated model. The truncated form of EXHALT is simpler than the full EXHALT model. Blue shooters and Blue effectiveness are input scalars that can be constants, time-indexed scripts, or time-dependent functions. A few Red advance configuration parameters are specified to calculate the size and rate of Red's advance. In the full model, many of the values that are inputs in the truncated version are derived from higher-

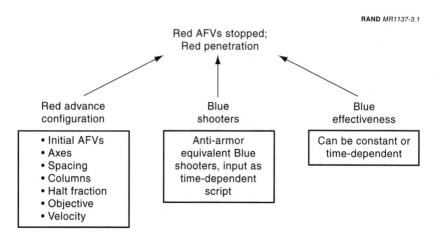

Figure 3.1—Structure Diagram of Truncated EXHALT

resolution calculations and indexed by the various Blue Shooter types. An example of this variable-resolution design is shown in Figure 3.2, which illustrates how the Blue shooters node can be expanded into a more-detailed calculation.[6] The boxes for Losses and the Commander Model indicate that those nodes represent higher-resolution calculations in themselves. These calculations, as well as those for Blue Effectiveness, can be similarly represented in treelike diagrams of their own.[7]

Moving between levels of resolution is not always easy or neat. In EXHALT, many parameters are affected by other parts of the model. Most of these interactions are explicit in the full version of EXHALT.[8] However, as EXHALT is truncated toward the simpler model illus-

---

[6]Underlined variables are vectors, indexed by Blue Shooter type, in the full EXHALT model.

[7]The tree diagrams for Losses, the Commander Model, and Blue Effectiveness are not shown here. Interested readers are directed to EXHALT itself for details regarding these modules.

[8]Although many interactions have been made explicit in EXHALT, some have not. As an example, a possible correlation between Red configuration and Blue effectiveness, in which Red's spacing affects the effectiveness of Blue's area weapons, is not implemented; however, increasing EXHALT's resolution by adding the weapons carried by Blue's shooters would explicitly define this relationship.

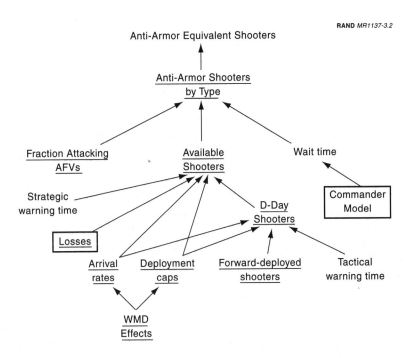

Figure 3.2—Illustrative Expansion of "Blue Shooters" Node

trated in Figure 3.1, the interactions are no longer modeled and must be taken into account implicitly (or at least kept in mind) when setting EXHALT's input values. This does not imply a failure of MRM but rather admits the hard truth that—in the real world—interactions exist between systems that can cause exact representations to be extremely complex. However, modelers can sometimes make approximations that do not significantly reduce the accuracy of their models but that greatly increase their insight-generating functionality.

When simpler model representations are needed (for the level of analysis desired or due to computer resource limitations), much of the complexity of the model can be avoided by this truncation (although much richness is lost, as well). Choice (or construction) of the appropriate level of EXHALT resolution will depend on the needs of each user. Fortunately, EXHALT's core structure, being MRM, is very flexible and adaptable.

EXHALT Outputs

The default output page of EXHALT is shown in Figure 3.3. While these output nodes represent many of the commonly used results that can be obtained from similar models describing the halt problem, EXHALT is able to utilize the Analytica modeling environment, which allows the user to view the results of any variable in the model by selecting the node of choice and clicking on the "Results" button. A brief tutorial on reading result windows in Analytica is given in Appendix C.

Another useful feature of Analytica and EXHALT is the ease with which users can create their own output nodes. For example, the "strength at defense line" measures were added after the rest of the model was complete; this simply involved creating a new node and entering the appropriate algorithms in terms of the model's other variables.

Viewing Results

EXHALT can be run in many ways: single-value runs, multiple lists for parametric exploration, uncertainty distributions for probabilistic exploration, or some combination of parametric and probabilistic exploration. One "standard" way to run the model is to replace many inputs with lists of values that span the problem space one wishes to examine and then to "scroll" through the inputs while viewing the result display. This parametric exploration can be extremely informative and convenient.

Using multiple lists can become extremely memory-intensive, depending on the number of lists and the number of values per variable. Hundreds of megabytes of RAM or virtual memory may be needed. However, colleague Manuel Carrillo developed for us a Perl script for doing "batch runs" with Analytica (Appendix D). This can generate data tables defined by multiple lists for the scenario space, which avoids the memory problem by running cases one at a time. The tool accumulates the results in a table that can be opened from Microsoft Excel™ or some other spreadsheet application. That data can also be used in UNIX-based systems for viewing parametric results.

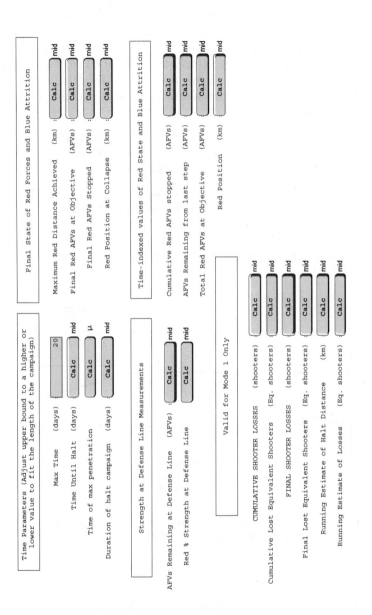

Figure 3.3—Standard Output Nodes in EXHALT

Alternatives are (Davis and Hillestad, forthcoming) (1) probabilistic exploration in which one uses probability distributions to represent uncertainty and (2) a combination of parametric and probabilistic exploration in which one uses some lists and some distributions. Either of these methods greatly reduces the memory burdens. Users are advised to view results in terms of mean values of cumulative probability distributions, not midvalues, for reasons explained in the Analytica documentation.[9]

In most of what follows, we will show results for deterministic parametric exploration, but in our analysis, we routinely use a combination of parametric and probabilistic exploration.

Typical results of interest will be halt time, maximum Red distance achieved, and Red strength at some specified interim point. To illustrate how the result displays are used, we show two examples. First, Figure 3.4 shows results for the halt time when the C⁴ISR system is varied and strategic warning time is varied parametrically, with values from 2 to 8 days.

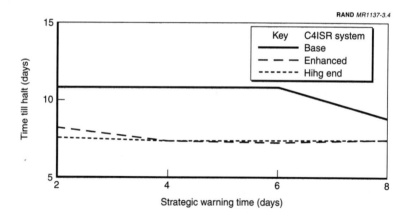

Figure 3.4—Result Window for the Time Until Halt Result, Varying C⁴ISR System and Strategic Warning Time

---

[9]Unless a function is linear (which EXHALT certainly is not), $f(\langle x \rangle) \neq \langle f(x) \rangle$. Thus, viewing the results of a model using only the midvalues of any specified probability distribution does not grant insight into the effects of uncertainty; on the contrary, it can be outright misleading.

Analytica can only display two input variables (plus the output) at a time on any particular chart, but the Analytica result window allows the user to "page" through many variables using scrolling lists, as shown in Figure 3.5. Note that, in addition to seeing the two curves for different Red column speeds, the user also has the option of "scrolling" through different values for the number of axes of Red's advance. Any number of variables can be explored in this way, but there are memory and run-time constraints.[10]

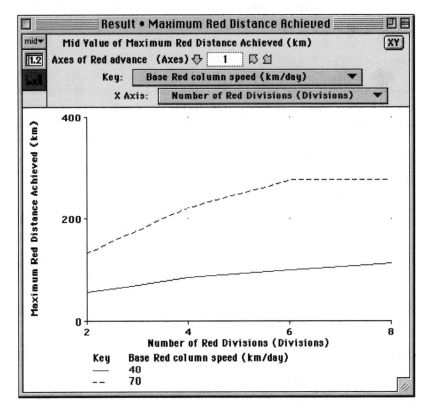

Figure 3.5—Result Window for "Maximum Red Distance Achieved"

---

[10]Another shortcoming is the relatively limited nature of Analytica's graphics package. However, the data can be exported to Excel or other programs, and a new version of Analytica (for the PC only) reportedly connects well with Excel's features.

Finally, the MRM design of EXHALT makes it easy to "walk through" the model, looking at the results of various parameters quickly to gain insight into which parameters are important and how they affect EXHALT's results. It can be very effective and efficient to define sets of variables of interest as multivalued lists or as probability distributions and then to use Analytica displays to view means, probability bands, cumulative distributions, or other statistics. Drilling down through levels of resolution is also useful in understanding parametric interactions and relationships within the EXHALT model.

Sample EXHALT Results

We can illustrate the utility of EXHALT with several examples, all of which are notional in construction. In the first case, we shall look at how part of the Terrain/Dispersion Effects module affects results. In this example, Blue is attempting to halt an advance comprised of five Red divisions, each with 1,000 AFVs. To simplify the analysis, Blue's wait time, during which he flies at a reduced rate out of concern for Red's air defenses, is fixed at four days.[11]

We will examine the effects of Red's threatening the use of WMD and/or mining on Blue's interdiction of Red's advance. To do this, we will set the "WMD/Mining Present?" yes-no choice to "All," causing Analytica to evaluate EXHALT for both values. In addition, we will parametrically vary Competence Time from Warning, which describes the increase in effectiveness as Blue gains competence with his C[4]ISR assets, and the C[4]ISR system used by Blue, evaluating

---

[11]Here, we "turn off" the Commander's wait-time decision agent. This simplifies discussion because, with the wait-time agent operating, there can be some nonintuitive (but heuristically reasonable) discontinuities and nonmonotonicities, as is common with decision models and real people. If, for example, strategic warning increases from four to six days, which one might expect to improve results, the wait-time agent may conclude that his prospects are good even if he waits longer than with only four days of warning before committing vulnerable aircraft. Although this will increase penetration distance, he may estimate that the distance-losses combination will still count as a "good" outcome. In contrast, with only four days of strategic warning, the wait-time agent may conclude that his prospects are at best "fair" and that he should commit his vulnerable aircraft very early. This may result in even less penetration than with greater strategic warning (but a less-good overall outcome). Such effects depend, of course, on the trade-off curves used (which are input).

for Base, Enhanced, and High-End systems. Other input parameters, which will be the same in both cases unless specified otherwise, are described in Table 3.2.

Given these parameters for exploration, EXHALT allows us to study their effects on any node in the model. Drawing on the default output nodes shown in Figure 3.3, we will view results for Time Until Halt, Maximum Red Distance Achieved, and Red's Position (over time).

A screen snapshot of an outcome table for Time Until Halt, which is the time it takes Blue to destroy the halt fraction of Red's AFVs, is shown in Figure 3.6. In this display, two of the input parameters are shown in tabular form, while the value of the third (and subsequent) input parameter is shown above. Thus, Figure 3.6 shows the Time Until Halt for various C^4ISR Systems and Competence Times from Warning, when the threat of WMD and/or mining is present. A similar table, showing the results over the same values of C^4ISR System and Competence Time, except with no threat of Red's using WMD and/or mining, is shown in Figure 3.7. (In Analytica, users would simply click on one of the arrows next to "WMD/Mining Present?" to select a different value for that parameter.)

Depending upon memory and run-time constraints, users can build "master-table" type displays, incorporating many list-defined variables, to explore in a table such as those in Figures 3.6 and 3.7, although the authors note that these requirements increase exponentially as new lists are added (Analytica must calculate permutations across all lists).

The Time Until Halt results in Figures 3.6 and 3.7, though, say nothing about how far Red advanced before being stopped or how Red's advance proceeded over time. EXHALT easily lets us explore these variables as well. Figure 3.8 is a screen shot of Red's maximum penetration distance, exploring these same variables, here displayed graphically. Note that, in all cases, Red's penetration distance is substantial (over 200 km, well into Saudi Arabia if the Halt scenario is Desert Storm–like). Interestingly, note that Blue's results are the same for Competence Times of three, four, and five days. This suggests that, given Blue's warning time of five days, Blue's being particularly fast at gaining competence with the use of his C^4ISR assets

Table 3.2

EXHALT Input Parameters for Text Example
(Notional Unclassified Values)

Blue Inputs	
D-Day shooters in theater	84 F-22s, 60 F-18s, 50 B-1s, 84 F-15s, 500 naval missiles
Arrival rate	In per-day shooters: 12 F-22s, 12 F-15s; 60 F-18s arrive 10 days after strategic warning time
Maximum deployment	144 F-22s, 180 F-18s, 50 B-1s, 144 F-15s, 500 naval missiles
Shooter capacity of theater	144 F-22s, 180 F-18s, 50 B-1s, 144 F-15s, 500 naval missiles
Nominal sorties or shots per day	F-22s: 2; F-18s: 2; B-1s: 0.5, F-15s: 2
Nominal kills per sortie or shot	F-22s: 2; F-18s: 3; B-1s: 12; F-15s: 3; naval missiles: 1 per missile
Fraction flying during SEAD	F-22s: 100%; F-18s: 10%; B-1s: 0%; F-15s: 50%
Anti-armor fraction	F-22s: 50%; F-18s: 50%; B-1s: 100%; F-15s: 50%; naval missiles: 50%
Initial shooter loss rate	F-22s: 1%; F-18s: 4%; B-1s: 10%; F-15s: 4%
SEAD time	5 days
Flexibility of fires	Moderate
C^4ISR system	Base, enhanced, and high end
Mode of Blue attack	Leading edge
WMD/mining?	Yes and no
Comp. time from warning	3, 4, 5, 6, 7 and 8 days
Access constraints?	No
Red Inputs	
Overall Mean AFV spacing	0.1 km
Base Red column speed	70 km/day
Red time concentration	Moderate
AFVs per division	1,000
Axes of Red advance	1
Columns per axis	2
Model Assumptions	
Tactical warning time	5 days
Tactical-strategic delta	0 days
Overall halt fraction	0.50
Unit break point	0.70
Distance to objective	600 km

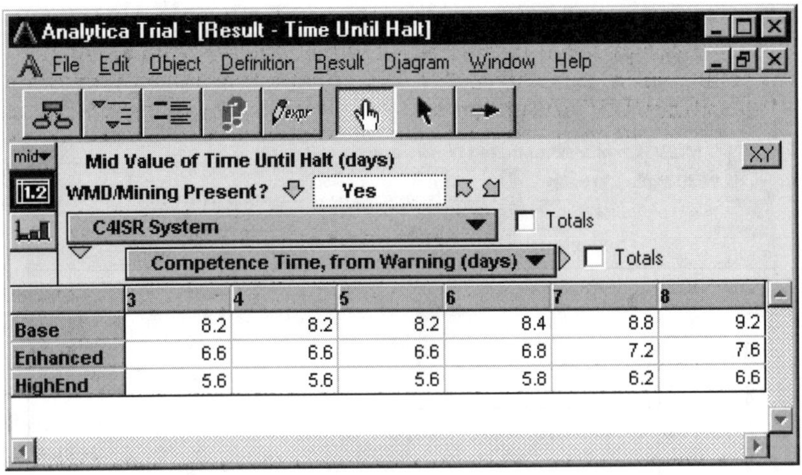

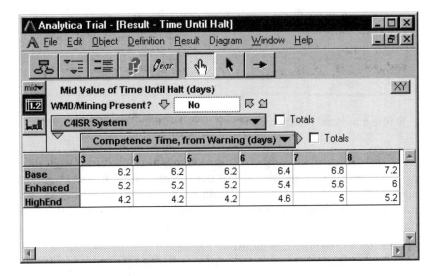

Figure 3.6—Time Until Halt Result with WMD/Mining Threat

Figure 3.7—Time Until Halt Result Without WMD/Mining Threat

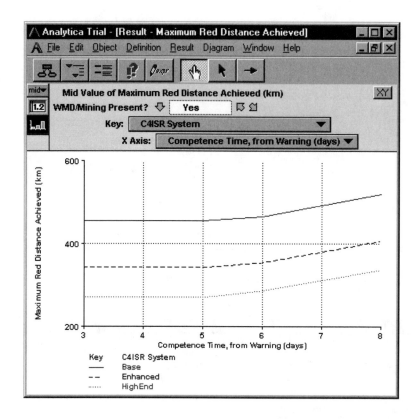

Figure 3.8—Analytica Screen Shot of Maximum Red Distance Achieved

does not necessarily help him. Although Blue certainly does not want to be slower at gaining C⁴ISR competence than five days, being able to do it in three days has no additional benefit. EXHALT allows for the exploration of these types of interactions that account for so much of the richness of the early interdiction problem.

Figure 3.9 illustrates this point in another way: by examining the progress of Red's advance over time. For the first few days of the campaign, each of the C⁴ISR Systems yields equally good (or equally poor) results. However, once the wait time has been reached, and Blue dramatically increases the number of sorties flown, the Enhanced and High-End C⁴ISR systems yield greater rollback effects and reach the halt point (at which Red's forward progress is stopped) sooner. These two effects lead to much better penetration distances.

Further, looking at the Red Position results over time in this manner yields insight into the phenomenology of the halt problem.

A second example will illustrate the use of probabilistic distributions as EXHALT input variables. In this scenario, we include the wait-time decision agent. We otherwise set the input parameters to be identical to those for the previous case, except for the parameters to be explored probabilistically (described below).

In examinations of the halt problem, studies often assume a deterministic mean velocity for Red's advance, as well as Red's overall halt fraction (the fraction of attrition Red can take before his advance

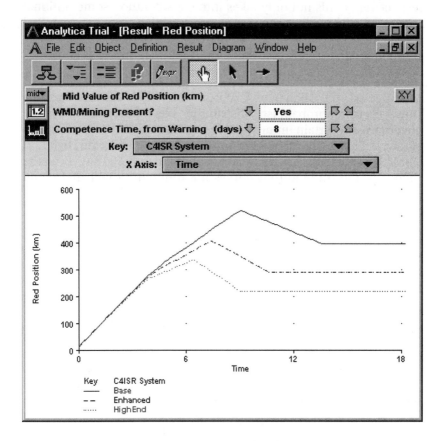

Figure 3.9—Analytica Screen Shot of Red Position Over Time

collapses). Often, however, it is useful to allow these parameters to vary randomly over some range, either to explore the outcome space of the model over a large domain or to reflect the uncertain nature of these parameters. In this case, for both Red's base column speed and Red's overall halt fraction, we assign triangular distributions. The triangular distribution is parameterized by three numbers: a minimum, a maximum, and a mode. The probability of the distribution taking a certain value increases linearly from the minimum to the mode, then decreases linearly to the maximum, as shown in Figure 3.10. Red's base column speed has a minimum of 40 km/day, a maximum of 80 km/day, and a mode of 60 km/day. Red's overall halt fraction has minimum, maximum, and mode of 0.2, 0.6, and 0.5, respectively. This not only takes into consideration some notional "mean value" but also accounts for conservative cases in which Red is "tough" and generous cases in which Red is easily broken. The cumulative distribution of Maximum Red Penetration is shown in Figure 3.11.

As expected, the High End C⁴ISR system yields the best results. However, using probabilistic distributions as inputs illustrates how uncertainty in certain input variables is reflected in the distributions of the result; e.g., the 90 percent confidence interval for the High End C⁴ISR System results has a spread of approximately 200 km. Even the shapes of the three output curves are different, a fact that could

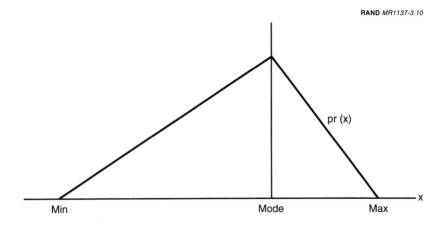

RAND MR1137-3.10

Figure 3.10—The Triangular Distribution

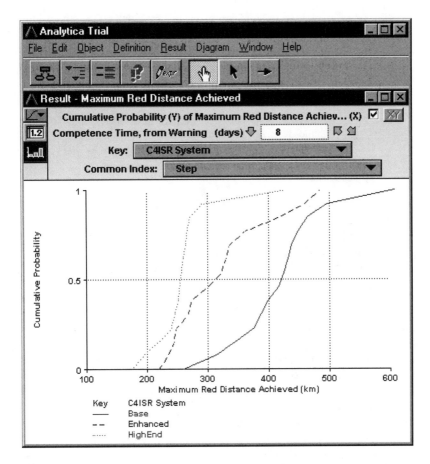

Figure 3.11—Cumulative Distribution of Maximum Red Distance Achieved

never be understood by looking at deterministic outputs. Finally, deterministic models can actually be deceptive in the simplicity of their results, never revealing the complex structure present in many systems. Compare the output in Table 3.3, which shows the deterministic calculations for Maximum Red Penetration using the medians of the distributions used for Figure 3.11 with the richness present in the cumulative probability chart.

Users need not be limited, however, to examining only the results we have identified as useful and placed in the "Outputs" module. An

Table 3.3

**Deterministic Calculation of Red
Penetration Distances**

C⁴ISR System	Maximum Red Penetration
Base	421 km
Enhanced	277 km
High End	261 km

NOTE: For Competence Time from Warning equal
to 8 days.

advantage that Analytica provides to EXHALT is the ability to observe easily the value(s) of any parameter in the model, be it input, output, or an intermediate calculation. Our last example illustrates this point. Instead of viewing the results of maximum distance achieved, or some other standard output measure, we will observe the result of a calculation "internal" to EXHALT: the day-by-day decision arrived at by the Blue wait-time decision agent.

For this illustration, we will use the same input parameters as in the previous example, except for the following: We only consider a Competence Time from Warning of six days; we remove the probabilistic nature of Overall Halt Fraction and Base Red Column Speed, using 0.5 and 70 km/day, respectively; and we observe the wait time Blue would choose with the Base C⁴ISR system. The wait-time decision results are shown in Figure 3.12. As a model assumption, Blue always begins the campaign at t = 0 assuming a wait time of 1 day (i.e., Blue begins the campaign in wait mode). After his first round of attacks, he observes his results, projects his losses and Red's penetration, and decides on his best wait time. From then on, at the beginning of each day, Blue may adjust his wait-time decision based on what he has learned. In the case shown in Figure 3.12, Blue adjusts his wait time to 6 days after the first round of attacks, then to 7 days at t = 1 day, to 8 days at t = 2 days, and temporarily back to 7 days at t = 6 days. Simulation time never reaches the wait time until t = 8, though, so Blue remains in wait mode until that time.

Increasing the number of Red divisions in the advance to eight changes Blue's wait-time calculation, with interesting results, shown in Figure 3.13. In this case, Blue's first round of attacks leads him

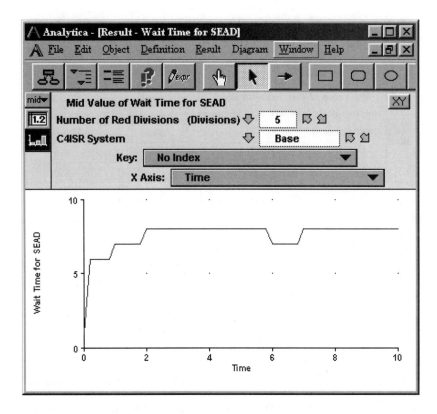

Figure 3.12—Wait-Time Decision Versus Time

initially to set his wait time to 3 days. As he learns about his effectiveness and Red's progress, however, he adjusts his wait time up and down through the course of the campaign, eventually settling on a wait time of 8 days after t = 6 days.[12] The effect that this has on Blue's wait mode can be observed more easily by creating a Test variable, which is 1 when Blue is in wait mode (i.e., when t < wait time) and 0 when Blue attacks full force. The Test variable is plotted in Figure 3.14.

---

[12]The wait-time decision agent selects the wait time used by Blue from a given set of "candidate" wait times. In this example, 8 days was the highest candidate wait time.

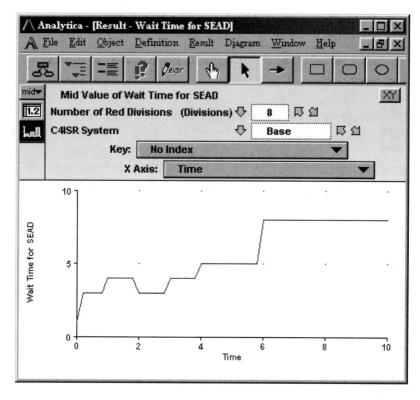

Figure 3.13—Wait-Time Decision Versus Time for Eight Red Divisions

Note that, in this case, Blue ends his wait at the beginning of Day 6 (t = 5) and attacks at full strength. However, Blue's daily projections results in his deciding to reenter the wait mode at t = 6, remaining there for two days before returning to his full-strength attacks. Thus, the use of a decision agent to set wait time has allowed Blue to be flexible in his strategy in response to his gaining information about the campaign.

CONCLUSIONS

In this chapter, we have described EXHALT's input parameters and worked through some examples of EXHALT results and modes of operation. EXHALT is a powerful and adaptable tool for use in

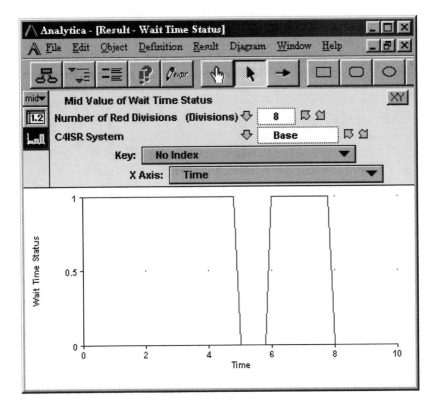

Figure 3.14—Test Variable Showing Blue's Wait Mode Versus Time

studying the halt problem across a wide range of scenarios and from many different perspectives. Its degrees of freedom may be explored parametrically and probabilistically and, perhaps most usefully, in any combination of the two. Unlike many similar models, EXHALT employs decision agents that allow Blue to adjust his force allocation strategy to respond to particular conditions within the scenario space. Finally, given Analytica's visual modeling environment and multiresolution design, EXHALT is easy to understand structurally and, importantly, is easy to adjust to add new features and resolution levels.

This concludes our overview of EXHALT, except for certain details provided in the appendices. Of course, the best way to understand EXHALT and how it can be used is to explore the model as it runs in the Analytica environment on the Macintosh or PC. As noted earlier, EXHALT was developed with the intention of being largely self-documented. As a practical matter, EXHALT will be modified as time goes by. Thus, although we expect this report to remain largely correct, the "definitive" documentation is and will be EXHALT itself.

IMPLEMENTING A DECISION AGENT WITHIN EXHALT: THE COMMANDER'S WAIT-TIME DECISION

This appendix provides a detailed look at how the Blue commander model (a "Blue agent") estimates halt time,[1] eventual Red penetration and Blue losses as a function of when he chooses to begin operating his aircraft at full sortie rates, and how he then decides on this "wait time."

ESTIMATING THE HALT TIME AND MAXIMUM RED DISTANCE ACHIEVED

The first step the Blue commander must take is to estimate the time at which Blue can halt Red's armored advance, as a function of wait time (the time Blue waits, because of Red's initially powerful air defenses, before flying full force). From this value, he can estimate eventual Red penetration and Blue losses. The commander model bases this estimate on information available from the model inputs and other observables during the course of the simulation.[2]

---

[1]In this report, "halt time" refers to when Red's advance "collapses" as the result of overall attrition. An alternative definition would focus on the time (the turnaround time) at which Red's forward edge ceases to advance. In the case of a Blue leading-edge strategy, this can be followed by a period in which Red's columns continue to move even though their noses are being rolled back. We plan to give the Blue commander the option to focus on the distance of maximum penetration rather than the penetration as of the time the columns collapse. This might be important if Blue were concerned about a defense line being overrun or some important position (e.g., a city) being occupied, even if temporarily.

[2]In the default settings of the model, the Commander uses correct values for his time to suppress Red's air defenses, his long-term effectiveness, and Red's break points.

The basic equation used to calculate this halt-time estimate (Davis and Carillo, 1997), from D-Day (time zero), is

$$
\begin{aligned}
\xi &= \int_0^{T_h} \delta F(s) A(s) ds \\
&= \int_0^{T_h} \delta F(s) \left[A(0) + \int_0^s R dq \right] ds \\
&= \int_0^{T_h} \delta F(s) \left[A(0) + Rs \right] ds,
\end{aligned}
\tag{A.1}
$$

where ξ is Blue's estimate of the number of Red AFVs he needs to kill with his shooters to achieve a halt; $F(s)$ is the fraction of Blue's shooters that are attacking Red AFVs as a function of time (in practice, this will have one value for the waiting period and another for the post-wait period); δ is Blue's mean effectiveness estimate; and $A(s)$ is the number of Blue shooters deployed as of time s. $A(s)$ is, in turn, equal to the number that are present on D-Day [$A(0)$] plus the number that have arrived up to time s; R is the estimated arrival rate of equivalent shooters, taken to be constant. $A(0)$, or D-Day shooters, equals the number of shooters in the theater initially, plus the number that deploy during the tactical warning time.

Learning

The Blue Agent "learns" by comparing each of his estimates, above, with observed results as of time t. In addition to current shooter levels and arrival rates, Blue also observes, as he goes along, how many Red AFVs he has killed up to the current time, and how far Red has progressed in his advance. To take advantage of this cumulative kills information, Blue splits the integral:

However, EXHALT allows the user to set incorrect values for these parameters, in which case the commander model makes corrections to these initial estimates on the basis of experiences as the campaign unfolds. For example, the Commander can update his estimate of his effectiveness as he observes his shooters' attacks on Red's AFVs.

$$\xi = \int_0^t \delta F(s)A(s)ds + \int_t^{T_h} \delta F(s)A(s)ds$$

$$= Pastkills + \int_t^{T_h} \delta F(s)A(s)ds,$$

(A.2)

where t is the current time in the simulation, and Pastkills is the observed value for total kills by Blue. The equation to be solved has now been reduced to

$$\xi = Pastkills + \int_t^{T_h} \delta F(s)\big[A(0) + Rs\big]ds.$$

(A.3)

A number of cases have to be treated separately. The value of F changes as Blue goes from pre-wait to post-wait mode; thus, depending upon whether the wait time falls within the time interval spanned by the integral in Equation A.3 and whether the current time in the simulation is in the pre- or post-Wait period, Equation A.3 will have to be solved in different ways. We will work out details for only some of them, since the other cases follow readily.

When the halt time occurs during the wait period or when the time in the simulation is after the wait period is over, the integral can be solved intact, since F remains constant in each of these cases:

$$\xi = Pastkills + \delta F\left[\left(A(0)T_h + \frac{R}{2}T_h^2\right) - \left(A(0)t + \frac{R}{2}t^2\right)\right].$$

(A.4)

To find the halt time, Equation A.4 can be rearranged into the form of a standard quadratic equation for T_h:

$$\frac{R}{2}T_h^2 + A(0)T_h - \left[A(0)t + \frac{R}{2}t^2 + \frac{\xi - Pastkills}{\delta F}\right] = 0,$$

(A.5)

using whichever F (pre- or post-wait) is appropriate. Solving via the quadratic formula,

$$x = \frac{-b \pm \sqrt{b^2 - 4ac}}{2a} \quad \text{if} \quad ax^2 + bx + c = 0,$$

(A.6)

one obtains (ignoring the negative root, which is nonphysical)

$$T_f = \frac{-A(0) + \sqrt{A(0)^2 + 2R\left(A(0)t + \frac{R}{2}t^2 + \frac{(\xi - \text{Pastkills})}{\delta F}\right)}}{R}. \tag{A.7}$$

However, when the current time in the simulation is during the wait period and when halt will occur after the wait period, the integral in Equation A.3 must be split, yielding

$$\xi = \text{Pastkills} + \int_t^{T_w} \delta F_{\text{pre}}\left[A(0) + Rs\right]ds \tag{A.8}$$

$$+ \int_{T_w}^{T_h} \delta F_{\text{post}}\left[A(0) + Rs\right]ds,$$

where T_w is the length of Blue's wait period, before which his shooters attack at a reduced rate out of concern for Red's air defenses, and F_{pre} and F_{post} are the estimated fraction of Blue's equivalent shooters attacking Red AFVs during the pre- and post-wait periods, respectively. Integration yields

$$\xi = \text{Pastkills} + \delta F_{\text{pre}}\left[A(0)(T_w - t) + \frac{R}{2}\left(T_w^2 - t^2\right)\right] \tag{A.9}$$

$$+ \delta F_{\text{post}}\left[A(0)(T_h - T_w) + \frac{R}{2}\left(T_h^2 - T_w^2\right)\right],$$

which can be rearranged into the form of the standard quadratic equation:

$$\frac{R}{2}T_h^2 + A(0)T_h + \frac{F_{\text{pre}}}{F_{\text{post}}}\left\{A(0)(T_w - t) + \frac{R}{2}\left(T_w^2 - t^2\right)\right\} \tag{A.10}$$

$$-\left\{A(0)T_w + \frac{R}{2}T_w^2\right\} - \frac{(\xi - \text{Pastkills})}{\delta F_{\text{post}}} = 0.$$

Equation A.10 is solved in the model using the quadratic formula (Equation A.6):

$$T_h = \frac{-A(0) + \sqrt{A(0)^2 - 2R\left(\dfrac{F_{pre}}{F_{post}} D\right)}}{R}, \tag{A.11}$$

where

$$\begin{aligned}
D &= A(0)\left(T_w - t\right) + \frac{R}{2}\left(T_w^2 - t^2\right) \\
&\quad - A(0)T_w + \frac{R}{2}T_w^2 - \frac{\left(\xi - \text{Pastkills}\right)}{\delta F_{post}}.
\end{aligned} \tag{A.12}$$

When the current time in the simulation is after the wait period is over and when halt has yet to be achieved, the halt time can be calculated exactly as described in Equations A.3 through A.7, except that F, in this case, is the post-wait value of the fraction of shooter attacking Red AFVs (instead of the pre-wait value used in Equations A.3–A.7).

Adjusting for the One-Time Arrival of a Group of Shooters

Since the arrival rate, R, is taken to be constant, the above equation is unable to account for the arrival of some number of shooters at one point in the future (e.g., a group of F-18s arriving with a second CVBG some time into the campaign). If such an event will occur, the above procedure will overestimate the time it takes Blue to halt Red's advance. If the Blue commander knows about this event, he can adjust his halt-time projection to compensate for his overestimation.

In this compensation, the unaccounted-for shooters are assumed to have no effect until the wait time is over. If these shooters arrive after the wait time has expired, this assumption introduces no error. Otherwise, this should result in an acceptably small error, since shooters fly at a reduced rate during the wait period and therefore have relatively little effect.

To understand the estimation process, and thus how we can correct for these arriving shooters, one can view the halt-time derivation as a process that calculates the number of days it will take to accumulate sufficient shooter-days, with shooters arriving at some rate and attacking with some effectiveness (measured in kills per shooter-day), to kill the required number of Red AFVs. If the arrival of shooters has been underestimated, the correction can be calculated from the incorrect halt-time estimate (which we will now call T_h'), the A(s) function used in the calculation of T_h' and information about the arrival of unaccounted-for shooters.

In the particular case in the model, a set of F-18s arrives in the theater with the second CVBG. The number arriving is known, as is the time at which the F-18s arrive. The assumption that none of the newly arriving F-18s contribute until after the wait time means that both the uncorrected and corrected cases will be the same up until the wait time. Afterward, in the uncorrected case (Figure A.1), the total number of shooter days accumulated in the post-wait period can be measured by calculating the area of the shaded portion of the graph. This can be written, using T_h' for the uncorrected halt time, as

$$
\begin{aligned}
\text{Shooter} - \text{days} = &\left[A(0) + RT_w\right]\left(T_h' - T_w\right) \\
&+ \frac{1}{2}\left[R\left(T_h' - T_w\right)\left(T_h' - T_w\right)\right].
\end{aligned}
\tag{A.13}
$$

The shooter-days accumulating in the "corrected" case, with the F-18s arriving with the second CVBG, are shown in Figure A.2. Note that, in addition to the light shaded area that is similar to the uncorrected case, shooter-days also accumulate in the dark shaded area from the F-18s that arrive at time T_{CVBG}. In terms of the corrected halt time, T_h, and with B representing the number of F-18s arriving with the second CVBG, expressed in equivalent shooters,

$$
\begin{aligned}
\text{Shooter} - \text{days} = &\left[A(0) + RT_w\right]\left(T_h - T_w\right) \\
&+ \frac{1}{2}\left[R\left(T_h - T_w\right)\left(T_h - T_w\right)\right] + B\left(T_h - T_{CVBG}\right).
\end{aligned}
\tag{A.14}
$$

Since the mean effectiveness and anti-armor fraction are the same in both cases (since we have treated the effects of new shooters as though they occurred after the wait time), the number of shooter-days needed to kill the required number of Red AFVs is the same as in the uncorrected case. Setting Equations A.13 and A.14 equal, one

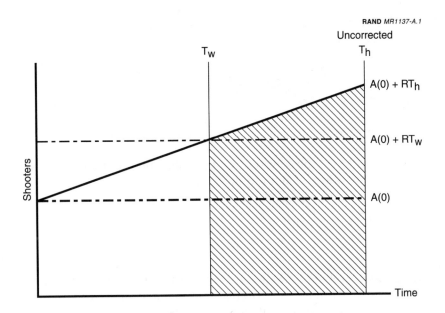

Figure A.1—Shooters Versus Time

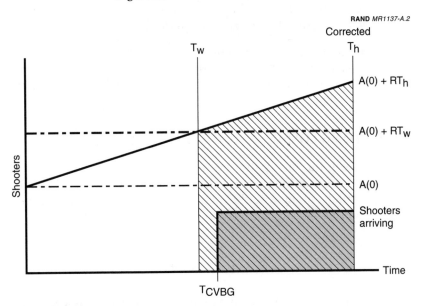

Figure A.2—Shooters Versus Time

can derive a quadratic equation of the standard form that can be solved for the corrected halt time:

$$\frac{R}{2}T_h^2 + \left[A(0) + B\right]T_h$$
$$- \left[\frac{R}{2}T_h'^2 + A(0)T_h' + BT_{CVBG}\right] = 0. \tag{A.15}$$

Equation A.15 can be solved using the following standard quadratic equation (ignoring the nonphysical negative root):

$$T_h = \frac{-\left(A(0) + B\right) + \sqrt{\left(A(0) + B\right)^2 - R^2 T_h'^2 + 2RA(0)T_h' + 2RBT_{CVBG}}}{R}. \tag{A.16}$$

With this corrected halt-time estimate, the Blue commander then proceeds to estimate the distance that Red's advance will achieve before it is halted, taking into account any projected rollback due to Blue's attacks. This distance estimate is updated during each round of attacks via the update of Blue's halt-time estimate, in addition to Blue's observing how far Red reached in the last step.

The Blue commander's estimate of Red distance achieved is forwarded to the Decision Table module for use in selecting the best wait time. His estimate for halt time is forwarded to the Estimate Losses module.

ESTIMATING LOSSES TO BLUE SHOOTERS

In estimating the losses that he will take for each candidate wait time, the Blue commander uses his estimates of the halt times for each candidate, along with the following expression for losses at any given time:

$$L(t) = S_{eq} L_0 F(t) A(t) e^{-\frac{2t}{T_{SEAD}}}, \tag{A.17}$$

where S_{eq} is the nominal sortie rate for the equivalent shooter type; L_0 is the loss rate per sortie or shot for the equivalent shooter type;

F(t) is the appropriate anti-armor fraction, depending upon whether t is in the pre- or post-waiting period range, A(t) is the number of equivalent shooters present at time t; and T_{SEAD} is the time it takes Blue to degrade Red's air defenses by a factor of 2e.

The equivalent loss rate per sortie or shot is derived from the loss rates of the separate shooter types and is calculated as the average loss rate, weighted by the number of shooters of each type and the sortie rates (relative to the sortie rate of the equivalent shooter) of each type, as follows:

$$L_0 = \sum_i \frac{L_{0,i} A(t)_{eq,i} S_i}{A(t) S_{eq}}, \tag{A.18}$$

where $L_{0,i}$ is the loss rate per sortie or shot for shooter type i; $A(t)_{eq,i}$ is the number of shooters of type i present at the current time, t, in the simulation (expressed in terms of equivalent shooters); S_i is the sortie rate for shooter type i; $A(t)$ is the number of total equivalent shooters in the theater at time t; and S_{eq} is the sortie rate of the equivalent shooter type.

While assumed to be constant for the purposes of estimating losses, the equivalent loss rate may change over time as new arrivals change the relative mix of shooters with different vulnerability. Additionally, Red's air defenses may change for unexpected reasons (additional assets are brought out of hiding, Blue's intelligence of Red's air defenses was faulty, etc.). Each round, Blue recalculates this loss rate by observing his losses in the previous time step and, inverting Equation A.17, solves for L_0:

$$\text{Updated\_}L_0 = \frac{L(t - timestep)}{S_{eq} F(t - timestep) A(t - timestep) e^{-\frac{2(t - timestep)}{T_{SEAD}}}}. \tag{A.19}$$

Blue also observes how many cumulative losses he has taken up to the current point in the simulation and, to estimate total losses, projects his future losses and adds them to those he has observed, as expressed as follows:

$$\text{Losses} = \text{PastLosses} + S_{eq}L_0 \int_t^{T_h} F(s)\Big[A(0) + Rs\Big] e^{-\frac{2s}{T_{SEAD}}} ds. \quad (A.20)$$

As with the calculation to estimate the halt time, the integral in Equation A.20 can only be solved intact if $F(s)$ is constant in the interval between the integration limits (i.e., if halt occurs during the waiting period or if the current time, t, in the simulation is after the end of the waiting period). Otherwise, the integral must be split just as before.

The commander's decision model solves Equation A.20 and sends the projected losses for each candidate wait period length to the Decision Tables module. There, it will be combined with the commander's projection of maximum Red distance achieved and input information about how these two factors are valued in Blue's success criteria to select the wait time that generates the most success for Blue.

THE COMMANDER'S DECISION CRITERIA

Once the Blue commander has estimated the maximum Red distance achieved and has projected his losses, he can execute the algorithm that will select the wait time that best meets his criteria. To do so, he uses a decision table that contains information about how he values certain benchmark combinations of maximum Red penetration and Blue losses.

The sample table shown as Figure A.3 is one that might apply to the *initial* territory taken by Red's advance. The entries in this table are input to represent the Blue commander's decision process. In this example, the table can be read as the Blue commander's stating, while looking at a map that might make distances meaningful in terms of borders, cities, and facilities to be defended even in a halt phase, "I would consider the following to be 'good' outputs: If I can stop Red within 284 km and take fewer than 15 losses; however, I would also be willing to take more losses, say up to 50, if I could stop Red within 142 km. The same goes for taking up to 70 losses, if I could stop Red within 80 km." The Blue commander goes on to define "fair" and "marginal" (and "poor" if he chooses) outcomes,

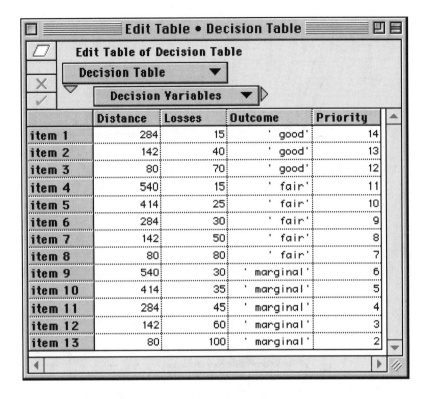

Figure A.3—Screen Shot of a Sample Decision Table

each with a unique ranking. Within each outcome (e.g., for each "good" outcome), the benchmarks were ranked on the basis of fewest Blue casualties being better. Note that, in this example, no outcome allowing Red to penetrate more than 284 km can be a "good" result for the Blue commander. This represents Blue's desire to halt Red far from important coastal facilities and oil fields.

The trade-off curves for this ambitious early halt case are shown graphically in Figure A.4. The benchmark distances correspond to key milestones between the Iraqi-Kuwait border and the Dhahran, Saudi Arabia. Figure A.5 graphically depicts these distances on a map of the region. Note that, in this example, no outcome that results in a Red penetration distance of greater than 280 km is con-

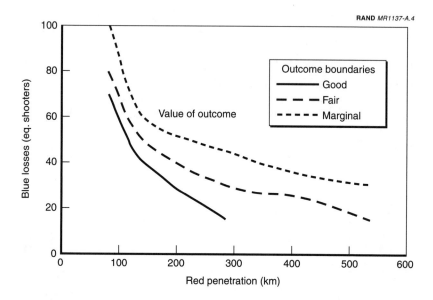

Figure A.4—Blue Commander's Indifference Curves

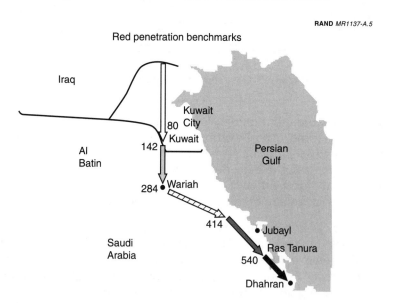

SOURCE: Modified from Davis and Carillo (1997).

Figure A.5—Distances in Persian Gulf Scenarios

sidered to be "good" by the Blue commander. This reflects Blue's desire to stop the Red advance at least by the time it reaches northern Saudi Arabia, far from important coastal facilities and oil fields.[3]

Once the wait-time decision criteria are entered to reflect the relative importance of stopping Red quickly versus minimizing casualties, the Blue commander can use his projections to choose his best wait time. To do this, Blue's estimates of his losses and the Red distance achieved for each wait-time candidate are compared to the appropriate columns of the decision table. An outcome matrix is built, indexed by wait-time candidate and benchmark item, with each entry in the table being the product of that benchmark item's priority; a scaling factor to ensure that, if two wait-time candidates match the same criteria, the longer wait time will have a slightly better score (since a longer wait time means fewer losses); and a "trigger" variable to indicate whether or not a given wait time met that benchmark's criteria. The decision module then selects the wait time that achieves the highest score in the outcome table. This process is repeated once each day in the simulation.

CONCLUSIONS AND A WORD OF CAUTION

We observe that the commander module appears to work well, although it may take some days before the commander's estimates stabilize. If he commits his vulnerable forces too early and takes high attrition, he will reinstitute the "wait." While the current model assumes Blue correctly knows his long-term effectiveness against Red's AFVs and the rate at which he degrades Red's air defenses, future work will generalize Blue's decision algorithm to allow the user to vary Blue's knowledge of the "correct" values for these and other parameters. Such work will allow the exploration of the effects of information and adaptiveness on Blue's halt efforts. We also intend to implement an option for the Blue commander to trade off the number of Red force reaching a specified defense line and the losses Blue takes to air defenses.

A closing word of caution: The authors note that this commander model was derived for the particular set of assumptions the authors

---

[3]The map in Figure A.5 and the benchmark distances and criteria were taken from Davis and Carrillo (1997).

used in applying EXHALT to particular analytic cases. If users significantly change EXHALT (as would be expected), they should carefully examine this appendix and the commander model module in EXHALT to ensure that the wait-time agent continues to be valid. In some cases, only trivial changes will be required (such as changing the name of a shooter type within the commander model to refer to a newly added missile type); in others, the commander's estimation algorithms may have to be rederived (if, for example, the arrival rate cannot be considered a constant). In any case, users can utilize this appendix to assist in determining the changes needed, and the wait-time decision agent can always be switched off until its validity has been verified.

INPUT DATA DICTIONARY FOR
THE FULL EXHALT MODEL

The following tables contain the complete lists of all input variables for the full version of EXHALT (i.e., Resolution Mode 1). Definitions and more-detailed explanations of these parameters, and all parameters input or calculated in EXHALT, can be found in the node documentation within the EXHALT model.

As described in the text, EXHALT includes an on-line model, RPEM, to assist the user in calculating reasonable multipliers to Blue's effectiveness due to terrain and dispersion factors. Table B.4 shows and defines the variables used in this calculation. Note that RPEM is not directly connected to EXHALT's calculations. Rather, it can be used to help the user arrive at the terrain-dispersion multipliers for each shooter type, which must then be user-input into the terrain-dispersion multiplier vector.

Table B.1

EXHALT Input Parameters for Red's Forces

Parameter	Description
Red divisions	Red divisions taking part in the advance.
Red AFVs per division	Red armored fighting vehicles in each division (may include tanks, APCs, etc.).
Axes of Red advance	Major axes on which the Red force advances.
Columns within each axis	Columns within each Red axis (e.g., Red AFVs may ride two across on a major road).
Base Red column speed	Absent the rollback from Blue's attacks, the number of kilometers per day traveled by the Red force.
Mean spacing between Red AFVs	The mean distance between a Red AFV and the AFV ahead of it, in meters.
WMD/mining flag	A flag to note if Red uses or threatens the use of WMD and/or mining of appropriate water-ways. If so, some Blue shooter types suffer effects to deployment rate, sortie rate, and theater capacity (due to safety measures or use of a lower-capacity alternate base).
Access constraints?	Indicates the ability of Red and/or his allies to constrain the ability of Blue to use air bases, staging areas, etc., convenient to the theater of war. Affects Blue's theater capacity.
Red time concentration factor	Red's ability to concentrate his movement during certain parts of the day (e.g., use "dash-and-hide" tactics).
Distance to objective	The mean distance, across Red's various axes of advance, from the starting point of Red's advance to his objective.

Table B.2

EXHALT Input Parameters for Blue's Forces

Parameter	Description
Shooter types	The types of shooters available to Blue. May be manned aircraft or missiles.
Shooters in theater initially	Shooters of each type forward-deployed to the theater.
Arrival rate	Shooters of each type arriving in the theater each day, once tactical warning has begun.
Maximum deployment	Maximum shooters of each type that may be deployed to the theater.
Shooter capacity of theater	Maximum shooters of each type that may be present in the theater at any time.
Nominal shots or sorties per day	Sorties or shots per day typically executed by each shooter type.
Nominal kills per shot or sortie	Kills achieved by each shooter type during a typical shot or sortie.
Fraction flying during wait period	Fraction of each shooter type that operates during wait period.
Nominal anti-armor fraction	Nominal fraction of each shooter type allocated to attacks on Red's AFVs.
Initial shooter loss rate	Loss rate on D-Day of each shooter type (per sortie or shot) due to Red's air defenses.
Mode of Blue attack	Blue's attack strategy against Red's AFVs. May be "leading edge" or "in depth."
SEAD time	Days required to suppress Red's air defenses by a factor of $2e$.
Wait time (input)	If the wait-time decision agent is turned off, this input parameter sets the length of time Blue will fly at a reduced rate out of concern for Red's air defenses.
Flexibility of fires	Blue's ability to adapt his attack to the movement patterns of Red.
CVBG arrival time	Days after strategic warning that a second CVBG arrives in the theater (with 60 F/A-18 E/Fs).
Competence time from warning	Days, from strategic warning, it takes Blue to build competence in the use of his C^4ISR assets.

Table B.2—Continued

Parameter	Description
C^4ISR system	From least effective to most effective, may be Base, Enhanced, or High End (assumed to be stealthy or space-based). Affects Blue's effectiveness.
Missile effectiveness threshold	Minimum level of Blue effectiveness for missile launch. A decision agent withholds missile attacks until effectiveness exceeds this threshold.
Wait-time decision criteria	A table setting out the decision criteria for the Wait-Time Decision Agent within EXHALT: halt distance, the losses the Blue commander would be willing to take to halt Red at that distance, and the utility of that mix of halt distance and losses to Blue.

Table B.3

EXHALT Input Parameters for Model Assumptions and System Variables

Parameter	Description
Unit break point	The fraction of forces in a front segment of Red's advance that Blue must kill to stop that segment of advance, if the leading-edge attack mode is used.
Overall halt fraction	The fraction of Red's overall force that Blue must stop to cause Red's advance to fall apart, or halt.
Strategic warning time	Days before D-Day during which Blue can take relatively risk-free and noncontroversial measures to prepare for a possible Red advance (e.g., deploy a second CVBG or "arsenal submarine" to the region).
Tactical warning time	Days for full-scale deployment before D-Day.
Maximum time	Days EXHALT will model. If the campaign is ongoing, but halt has not been achieved by the maximum time, the user should increase this value until halt is achieved. However, arbitrarily high numbers will result in needlessly high memory use.
Time step	The length of a time step in the model. May be set to achieve the desired time resolution.
Interim point	Some distance between the advance start point and the objective at which the user would like to observe how many AFVs remain (in terms of numbers and fractions).

Table B.4

EXHALT Input Parameters for Terrain-Dispersion Effects

Parameter	Description
AFVs per packet	In the terrain-dispersion calculation, Red AFVs are presumed to travel in packets. This is the number of AFVs in each packet.
AFV spacing within packet	Within each packet, the mean spacing between AFVs. This may be greater or less than the overall mean AFV spacing.
Mean spacing between packets	The mean tail-to-head spacing between AFV packets.
Mean AFV speed through open areas	Speed with which the packet moves through the open areas. May be greater or less than the average speed of movement and could be dictated by tactical countermeasures.
Speed estimation error	The fractional error in Blue's estimate of Red's velocity. Used to determine the standard deviation of the error in Blue's missile arrival time.
Mean length of open areas	The mean length of open areas in the otherwise closed terrain. Red AFVs may only be targeted and hit in open areas.
Time since last update	The time since Blue was last able to update his missile's targeting system with the packet's position and estimated velocity. It is over this time that Blue's speed estimation error will propagate.
Weapon footprint	The length of the footprint along the axis of the road going through the open area.
Maximum kills per missile	Self-explanatory. Presumably due to submunition limits, etc.
Fraction of targeted AFVs killed	The fraction of Red AFVs within the footprint of Blue's missile (and in the open) that will be killed.

VERIFICATION AND VALIDATION

No formal effort has been made to verify or validate EXHALT, since it is a relatively simple desktop model akin to more ubiquitous spreadsheet models. Many of the expressions in EXHALT that are used to estimate various real-life phenomena have not been formally reviewed for faithfulness but have been based on numerous detailed analyses at RAND and elsewhere that suggest these are reasonable representations. Other expressions are merely notional and are intended to provide rough estimates of various effects as placeholders until more phenomenological models can be developed. However, we have taken quite a number of quality-control measures, many of which exploit modern desktop-computer technology to address issues that we have long considered important to verification and validation[1]:

- *Conceptual model.* This report documents the conceptual model, as distinct from details of the implementing program.

- *Visual design.* The visual-programming methods of Analytica are extremely useful in sharpening issues of design and facilitating review. We made countless changes along the way as the result

---

[1]Methods for verification, validation, and accreditation are discussed in detail at the web site of the Defense Modeling and Simulation Office (http://www.dmso.mil/dmso.docslib). For a good review article, see Pace (1998). One of us (Davis) has long had an interest in verification and validation, especially technology for enhancing it (Davis, 1992). The Military Operations Research Society held a minisymposium on verification and validation methods at Johns Hopkins University's Applied Physics Laboratory in January 1999. The report of SIMVAL '99 is available on the web at http://www.mors.org.

of discussions focused on the diagrammatic representation of the model.

- *Verbose variable naming.* We have used relatively long and descriptive variable names to clarify meaning.

- *Array methods.* We have exploited a number of Analytica's array features, which greatly simplified mathematical expressions and, in the process, improved clarity and reviewability of the algorithms,

- *Automatic verification methods.* Analytica highlights variables it does not recognize (usually due to typographical errors) rather than assuming them to be new legitimate variables. In some cases, we replaced long If-Then-Else statements with tables, which greatly simplified review.[2]

- *Module by module testing.* We did extensive module-by-module testing, which is straightforward in the Analytica environment.

- *MRM.* Because of the MRM design, we were able to start module-by-module testing with highly aggregated modules and, once they appeared valid, to test submodels within the modules.

- *Visual displays of outputs.* All such testing is greatly assisted by graphical outputs, which can be generated on a node-by-node basis

- *Comparisons.* We have made numerous comparisons of results with those of previously developed models, including an analytic closed-form model for special cases, an Excel model we had used in earlier work, an Excel model developed by colleagues, and work done with a theater-level model (JICM).

- *Calibrations.* Some parts of EXHALT are calibrated to a more detailed model (PEM), which in turn was motivated by and calibrated to high-resolution simulation at the entity level (Davis, Bigelow, and McEver, forthcoming).

---

[2]The value of tables in the language itself was first dramatized in the RAND-ABEL language developed by Edward Hall and Norman Shapiro and used in the RAND Strategy Assessment System (RSAS). Our colleague Manuel Carrillo implemented the table feature for use in Analytica.

Although errors and other shortcomings will surely be found and although we plan to correct and improve standard versions of EXHALT from time to time as necessary, we believe that users can largely focus their concerns on (1) what assumptions to make about inputs and their uncertainties and (2) the inherent limitations of any model, such as EXHALT, that omits much of the richness of war (e.g., the role of maneuver forces, details of terrain, and operations with allies) to bring out certain analytically useful insights.

BATCH RUNS IN EXHALT

Given the memory-intensive nature of running multiple lists in Analytica, it is sometimes difficult to explore many parameters simultaneously in EXHALT. To get around this problem, our colleague Manuel Carrillo developed a Perl script that allows multidimensional parametric exploration of Analytica models by executing many single-value runs in a batch.

Each line in this script prompts Analytica to set specified parameters, execute the model, and then export specified results to a text file that can be read into a spreadsheet program, such as Excel, for analysis. The script then moves on to the next line, resetting values appropriately. To parametrically explore a multidimensional domain, the script should include a line for every permutation of input variables in the domain.

FORMAT

The format of the batch-run script is a text file, with each line in the following form:

    set|param1:=val1#param2:=val2#...#paramn:=valn|get|out1#out2...

Although the usual way to run the script is to set input parameters as scalars, values can also be set to text or arrays, but the form of the value (to be placed in the "param" placeholder in the script) must be identical to the form Analytica uses in the definition field when the definition is displayed in "expression mode."

If an output parameter is an array rather than a scalar, it must be preceded by '0+' in the script; i.e., out1 becomes 0+out1.

It is possible to write a simple Microsoft Excel module to generate a script automatically for all permutations of a scenario space, given the parameters involved and their individual domains.

PLATFORMS

Carrillo's Perl script takes advantage of Apple Events and is therefore only available on the Macintosh platform. Something similar can probably be developed easily for Windows platforms, particularly with Analytica 2.0, but we have not explored this as yet. In the Macintosh environment, we have used the script for batch runs consisting of over 20,000 iterations run over two days.

REVIEWING AND MODIFYING ALGORITHMS

As emphasized at the outset of the report, it is not our purpose here to review the Analytica modeling system or its language. EXHALT *is* a computer program; to understand it fully, much less to change it, one must understand both modeling and programming. That said, working with an Analytica model is much easier than programming in C or some other general-purpose language. To illustrate, suppose we wish to understand how the various situational multipliers are applied. If we open the EXHALT program by double clicking on its file, we obtain Figure E.1. If we then double click on the "EXHALT

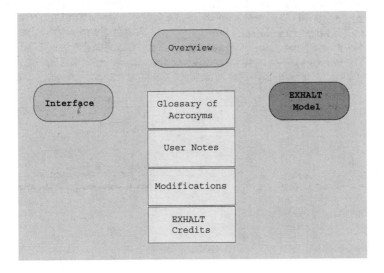

Figure E.1—EXHALT's Top-Level Window

Model" module, we obtain Figure E.2. When we began building EXHALT, we did so by constructing Figure E.2—without specifying details of the individual models.

If we now double click on the Blue Effectiveness module, we obtain Figure E.3. Again, drawing this figure was the way we designed the model. The figure is not after-the-fact documentation, but rather the starting point.

If we double click on the Kills per Shooter-Day module, we obtain Figure E.4. We are now at the lowest level in the model.

The window indicated in Figure E.4 shows two variables; the one on the left is the kills per shooter day (yes, this variable has the same name as the module within which it sits); the one on the right is a variant of the first expressed as a multiple of what a "standard shooter" (an F-15E) can kill in a day. Let us look at the former. If we double click on Kills Per Shooter-Day, we obtain the window shown in Figure E.5. This defines everything there is to know about the variable.

At the top of the window we see the computer's name for the variable. Below that we see the equivalent English-language title we have given it. It is this title that appears in the various diagrams. Reading downward, we see "description." This is an example of self-

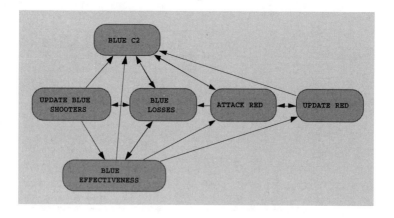

Figure E.2—The Halt Model

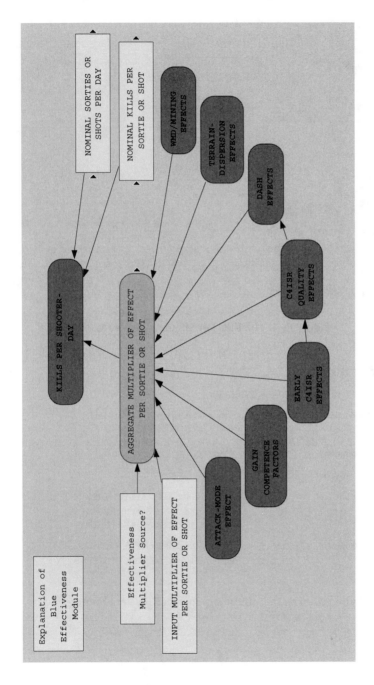

Figure E.3—The Blue Effectiveness Model

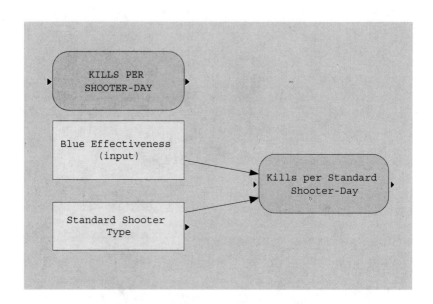

Figure E.4—The Kills Per Shooter-Day Module

○ **Variable** ▼	Killspershooterday	**Units:** AFVs/shooter-day

Title: KILLS PER SHOOTER-DAY

Description: Kills per shooter-day achievable by each aircraft type.

expr ▼

Definition: Effectiveness_mult*Kps*Sr

Value: [**Calc**   ]

Inputs: ○ Effectivenes...   AGGREGATE MULTIPLIER OF EFFECT PER SORTIE OR SHOT
              □ Kps   NOMINAL KILLS PER SORTIE OR SHOT
              □ Sr   NOMINAL SORTIES OR SHOTS PER DAY

Outputs: ○ Cum_eq_los...   Cumulative Lost Equivalent Shooters
              ○ Eq_aa_shoot...   EQUIVALENT ANTI-ARMOR SHOOTERS BY TYPE

Figure E.5—The Definition Window for Kills Per Shooter-Day

documentation: Whenever we introduce a new node, Analytica encourages us to fill in this description and thus provide an English-language explanation of what the node "means." Analytica, of course, cannot force us to do a good job.

Continuing, we find the "definition." This is the algorithm itself. It is equivalent to an equation saying:

$$\text{Killspershooterday}=\text{Effectiveness\_mult*Kps*Sr.} \qquad \text{(E.1)}$$

This, then, is the level of the model's "meat," where the algorithms are expressed. To be sure, we chose one of the easiest-to-understand nodes in the entire model, but it is sufficient to explain the visual-modeling environment.

Below the definition field are a list of inputs and a list of outputs (a misnomer, since it is actually a list of the variables that use Killspershooterday as an input). These are generated and updated automatically whenever we build the model visually by drawing arrows between nodes. That is, when we first built the model, we drew a diagram. Once the diagram existed, it implied that, when we got around to writing the equation defining Killspershooterday, its definition window (Figure E.5) would already show the inputs to that variable and the variables that use Killspershooterday as inputs. This makes it easier to write the algorithm in the first place because the correctly spelled computer names are available inside the window on menus and one merely has to add plus signs, multiplication signs, or whatever. In this case, the variable is just a product of Effective-ness_mult, Kps, and Sr.

We know that Kps and Sr are primitive inputs because they are indi-cated by rectangles. But what is Effectiveness_mult? If we double click on it, we obtain Figure E.6. Here we see that the overall multi-plier of effectiveness, relative to just Kps*Sr, is a product of a number of individual complicating factors, as discussed in the text. We see corrections for the dash tactic (with the correction depending on which employment strategy Blue is using), a factor indicating the level of risk the pilots are taking, and so on.

For the purposes of this report, let it suffice to note that we could trivially change this equation if we wanted to do so. We would

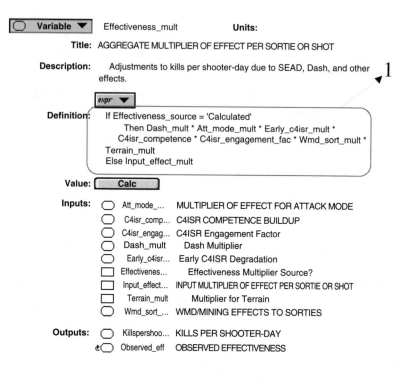

| ◯ Variable ▼ | Effectiveness_mult | **Units:** |

Title: AGGREGATE MULTIPLIER OF EFFECT PER SORTIE OR SHOT

Description: Adjustments to kills per shooter-day due to SEAD, Dash, and other effects.

expr ▼

Definition: If Effectiveness_source = 'Calculated'
Then Dash_mult * Att_mode_mult * Early_c4isr_mult *
C4isr_competence * C4isr_engagement_fac * Wmd_sort_mult *
Terrain_mult
Else Input_effect_mult

Value: [**Calc**]

Inputs: ◯ Att_mode_...   MULTIPLIER OF EFFECT FOR ATTACK MODE
◯ C4isr_comp...   C4ISR COMPETENCE BUILDUP
◯ C4isr_engag...   C4ISR Engagement Factor
◯ Dash_mult   Dash Multiplier
◯ Early_c4isr...   Early C4ISR Degradation
☐ Effectivenes...   Effectiveness Multiplier Source?
☐ Input_effect...   INPUT MULTIPLIER OF EFFECT PER SORTIE OR SHOT
☐ Terrain_mult   Multiplier for Terrain
◯ Wmd_sort_...   WMD/MINING EFFECTS TO SORTIES

Outputs: ◯ Killspershoo...   KILLS PER SHOOTER-DAY
ₑ◯ Observed_eff   OBSERVED EFFECTIVENESS

Figure E.6—The Definition for Effectiveness_mult

merely edit it. Or we could go back to the diagrams and indicate what additional variables should affect Effectiveness_mult, then return to this window to edit. That would have the advantage of making the additional variables be present on menus when we edit the equation.

Suppose, now, that we want instead to simplify. Suppose we want to "turn the multipliers off." All we have to do is replace the definition with "1." Further, we know that the *only* effect this will have in the entire model is to change the calculation of Killspershooterday, because that is the only variable that depends on Effectiveness_mult (see the list of outputs at the bottom). We could have seen this also by looking at the diagrams and observing that the only arrow out of Effectiveness_mult goes to Killspershooterday. In any case, as soon as we edit the definition, changing it to 1, the change takes effect.

There is no need to recompile. Instead, as in a spreadsheet, changes are interactive. Another useful feature is that, if one changes the names of nodes—either to sharpen up meaning or, during the design phase, as one thinks more clearly—editing a name anywhere causes the changes to ripple through the entire program automatically.

This should suffice to indicate some of the convenience of the visual-modeling environment, but let us again emphasize that most of the node definitions are much more complex and definitely look like "computer code." No one should expect to understand the inner workings of EXHALT fully, much less change it, without having a reasonable understanding of programming (e.g., at the level of BASIC). Moreover, much of the power of Analytica involves its use of mathematical arrays. While the arrays greatly simplify the equations and overall cognitive complexity for those who understand arrays, such arrays can be subtle and difficult for those who do not. Our advice to users is to go through Analytica's excellent tutorial, which uses live models, before attempting to understand fully or to change EXHALT. Even two days of effort can be very useful here.

BIBLIOGRAPHY

Belldina, Jeremy S., Henry A. Neimeier, Karen W. Pullen, and Richard C. Tepel, "An Application of the Dynamic C4ISR Analytic Performance Evaluation (CAPE) Model," McLean, Va.: MITRE, 1997.

Davis, Paul K., *Generalizing Concepts and Methods of Verification, Validation, and Accreditation (VV&A) for Military Simulations*, Santa Monica, Calif.: RAND, R-4249-ACQ, 1992.

Davis, Paul K., and James Bigelow, *Experiments in Multiresolution Modeling*, Santa Monica, Calif.: RAND, MR-1004-DARPA, 1998.

Davis, Paul K., James Bigelow, and Jimmie McEver, *Analytical Methods for Studies and Experiments on "Transforming the Force,"* Santa Monica, Calif.: RAND DB-278-OSD, 1999.

_____, *Effects of Terrain, Maneuver Tactics, and C4ISR on the Effectiveness of Long Range Precision Fires: A Stochastic Multiresolution Model (PEM) Calibrated to High-Resolution Simulation*, Santa Monica, Calif.: RAND, 2000.

Davis, Paul K., and Manuel Carrillo, *Exploratory Analysis of "the Halt Problem": A Briefing on Methods and Initial Insights*, Santa Monica, Calif.: RAND, DB-232-OSD, Santa Monica, CA, 1997.

Davis, Paul K., and Richard Hillestad, *Exploratory Analysis of Strategy Problems with Massive Uncertainty*, Santa Monica, Calif.: RAND, forthcoming.

Davis, Paul K., William Schwabe, Bruce Narduli, and Richard Nordin, *Mitigating Effects of Access Problems in Persian Gulf Contingencies*, Santa Monica, Calif.: RAND, MR-915-OSD, forthcoming.

Defense Science Board, *1998 Summer Study Task Force on Joint Operations Superiority in the 21st Century: Integrating Capabilities Underwriting Joint Vision 2010 and Beyond*, two volumes, Washington, D.C.: Office of the Under Secretary of Defense for Acquisition, 1998.

Gritton, Eugene, Paul K. Davis, Randall Steeb, and John Matsumura, *Ground Forces for a Rapidly Employable Joint Task Force: First Week Capabilities for Short-Warning Conflicts*, Santa Monica, Calif.: RAND, MR-1152-OSD/A, 2000.

Joint Chiefs of Staff, *Concept for Future Joint Operations: Expanding Joint Vision 2010*, Fort Monroe, Va.: Joint Warfighting Center, 1997.

Lumina Decision Systems Inc., *Analytica User Guide*, Los Gatos, Calif., 1996.

_____, *Tutorial: An Introduction to Analytica*, Denver, Co., 1997.

Ochmanek, David, Glenn A. Kent, Alex Hou, Ernst Isensee, Robert E. Mullins, and Carl Rhodes, *Joint Interdiction Capabilities Assessment: Interim Results*, Santa Monica, Calif.: RAND, unpublished manuscript.

Ochmanek, David, Edward Harshberger, David Thaler, and Glenn Kent, *To Find and Not to Yield: How Advances in Information and Firepower Can Transform Theater Warfare*, Santa Monica, Calif.: RAND, MR-958-AF, 1998.

Pace, Dale, "Verification, Validation, and Accreditation," in David J. Cloud and Larry B. Rainey (eds.), *Applied Modeling and Simulation: An Integrated Approach to Development and Operation*, New York: McGraw Hill, 1998.